Pocket AUTOMOTIVE DICTIONARY

William H. Crouse
Donald L. Anglin

Gregg Division
McGraw-Hill Book Company

New York
St. Louis
Dallas
San Francisco
Auckland
Düsseldorf
Johannesburg
Kuala Lumpur
London
Mexico
Montreal
New Delhi
Panama
Paris
São Paulo
Singapore
Sydney
Tokyo
Toronto

Library of Congress Cataloging in Publication Data

Crouse, William Harry, (date)
  Pocket automotive dictionary.

  1. Automobiles—Dictionaries. I. Anglin,
Donald L., joint author. II. Title.
TL9.C77   629.2'03   76-21820

ISBN 0-07-014752-3

Pocket Automotive Dictionary

Copyright © 1976 by McGraw-Hill, Inc. All rights reserved. Printed in the United States of America. No part of this publication may be reproduced, stored in a retrieval system, or transmitted, in any form or by any means, electronic, mechanical, photocopying, recording, or otherwise, without the prior written permission of the publisher.

ISBN 0-07-014752-3

1234567890 KPKP 7832109876

Cover photo taken by Dennis G. Purdy

The editors for this book were Ardelle Cleverdon and Myrna Breskin, the designer was Dennis G. Purdy, the art supervisor was George T. Resch, and the production supervisor was Rena Shindelman. It was set in Helvetica by York Graphic Services.
Printed and bound by Kingsport Press, Inc.

In This Book

The authors have produced many books in the automotive field, notably *Automotive Mechanics, Automotive Engines, Automotive Electrical Equipment,* among others. There are now twelve of these automotive books, covering all phases of the automotive service business. Each of the books has a glossary defining the automotive terms used in the book. Nowhere, until now, however, has there been one publication bringing together all of the automotive terms in common use in the service field.

There has been a demand for such a publication for a long time from teachers and students in the fields of automotive technology and automotive mechanics. The need has intensified in recent years with the increasing complexity of the automotive vehicle and with the introduction of much new technology such as electronics, solid-state physics, and modern emission control systems.

In compiling the dictionary, the authors have tried to include the automotive terms in common use in all facets of automotive service and performance. Thus, you will find such terms as "rod big end," "oversteer," and "throw a rod," defined in the dictionary. In recognition of the newer technological developments, the dictionary includes such terms as "monolithic timing," "A.C.I.D.," "PROCO," and "stratified charge." The performance field is also included with such terms as "Christmas tree," "gymkhana," "bad car," and so on.

No attempt has been made to produce a "comprehensive" dictionary of automotive terms and include those used in design and manufacturing. To do so would have made the dictionary much larger and more expensive. Our aim was to produce a dictionary that would be handy and useful to everyone in the automotive service business.

Definitions are only partial; that is, the definitions are not inclusive but cover only what specifically applies to the automotive service field. For instance, "apexes" is defined as the peaks on the rotor of a Wankel engine. "Air horn" is defined as a tubular passage in a carburetor.

A word or term in **bold face** within a definition means that the word or term is defined elsewhere in the dictionary. For example, "**low side** Same as **suction side**." indicates that **suction side** will be found elsewhere in the dictionary.

At the back of this book are abbreviations used in the U.S. Customary System and the SI System as well as approximate conversions from one system to the other. These conversions should help familiarize students with metric terms now being used in the automotive field.

The authors lay no claim to being lexicographers. Their aim was to produce a handy and useful dictionary that fulfills the needs of the automotive educators and students in the automotive service field.

William H. Crouse

Donald L. Anglin

**Other Books and Instructional Materials
by William H. Crouse and *Donald L. Anglin**

Automotive Chassis and Body*
  Workbook for Automotive Chassis and Body*
  Automotive Electrical Equipment
  Workbook for Automotive Electrical Equipment*
Automotive Engines*
  Workbook for Automotive Engines*
Automotive Fuel, Lubricating, and Cooling Systems*
  Workbook for Automotive Fuel, Lubricating, and Cooling Systems*
Automotive Transmissions and Power Trains*
  Workbook for Automotive Transmissions and Power Trains*
Automotive Service Business: Operation and Management
Automotive Emission Control
Automotive Engine Design
Workbook for Automotive Service and Trouble Diagnosis
Workbook for Automotive Tools
Automotive Mechanics
  Study Guide for Automotive Mechanics
  Testbook for Automotive Mechanics*
  Workbook for Automotive Mechanics*
  Automotive Troubleshooting Cards
The Auto Book
  Auto Shop Workbook*
  Auto Study Guide
  Auto Test Book*
  Auto Cassette Series
General Power Mechanics (with Robert Worthington and Morton Margules*)
Small Engines: Operation and Maintenance
  Workbook for Small Engines: Operation and Maintenance

**Automotive Room Chart Series
by William H. Crouse**

Automotive Electrical Equipment Charts
Automotive Engines Charts
Automotive Fuel Systems Charts
Automotive Emission Controls Charts
Automotive Engine Cooling Systems, Heating, and Air Conditioning Charts
Automotive Suspension, Steering, and Tires Charts
Automotive Transmissions and Power Trains Charts
Automotive Brakes Charts

**Automotive Transparencies
by William H. Crouse and Jay D. Helsel**
Automotive Brakes
Automotive Electrical Systems
Automotive Engine Systems
Automotive Transmissions and Power Trains
Automotive Steering Systems
Automotive Suspension Systems
Engines and Fuel Systems

ABOUT THE AUTHORS

William H. Crouse
Behind William H. Crouse's clear technical writing is a background of sound mechanical engineering training as well as a variety of practical industrial experience. After finishing high school, he spent a year working in a tinplate mill. Summers, while still in school, he worked in General Motors plants, and for three years he worked in the Delco-Remy Division shops. Later he became Director of Field Education in the Delco-Remy Division of General Motors Corporation for which he prepared service bulletins and educational literature.

During the war years, he wrote a number of technical manuals for the Armed Forces. After the war, he became Editor of Technical Education Books for the McGraw-Hill Book Company. He has contributed numerous articles to automotive and engineering magazines and has written many outstanding books about science and technology. He was the first Editor-in-Chief of the 15-volume McGraw-Hill Encyclopedia of Science and Technology. In addition, he has authored more than fifty technical books including *Automotive Mechanics,* which has sold over a million copies. His books have been widely translated and used in automotive mechanics training throughout the world.

William H. Crouse's outstanding work in the automotive field has earned for him membership in the Society of Automotive Engineers and in the American Society of Engineering Education.

Donald L. Anglin

Trained in the automotive and diesel service field, Donald L. Anglin has worked both as a mechanic and as a service manager. He has taught automotive courses in high school, trade schools, community colleges, and universities. He has also worked as curriculum supervisor and school administrator for an automotive trade school. Interested in all types of vehicle performance, he has served as a racing-car mechanic and as a consultant to truck fleets on maintenance problems.

Currently he serves as editorial assistant to William H. Crouse, visiting automotive instructors and service shops. Together they have coauthored magazine articles on automotive education and several books in the McGraw-Hill Automotive Technology Series.

Donald L. Anglin is a Certified General Automotive Mechanic and holds many other licenses and certificates in heavy duty truck mechanics, automotive education, service, and related areas. His work in the automotive service field has earned for him membership in the American Society of Mechanical Engineers and the Society of Automotive Engineers. In addition, he is an automotive instructor at Piedmont Virginia Community College, Charlottesville, Virginia.

A

A Abbreviation for **ampere.**

ABDC Abbreviation for *after bottom dead center;* any position of the piston between bottom dead center and top dead center, on the upward stroke.

abrasive A substance used for cutting, grinding, lapping, or polishing metals.

absolute pressure A pressure measured on a scale having, as its zero point, the complete absence of pressure (known as a *perfect vacuum*). Atmospheric pressure on the absolute scale is 14.7 psi [1.03 kg/cm² (kilograms per square centimeter)] or 29.92 in [760 mm (millimeters)] of mercury (Hg).

absolute zero The temperature indicating the complete absence of heat on the absolute temperature scale; equivalent to $-460°F$ $[-273°C]$.

A/C Abbreviation for **air conditioning.**

AC or **ac** Abbreviation for **alternating current.**

acceleration An increase in velocity or speed.

accelerator A foot-operated pedal, linked to the throttle valve in the carburetor; used to control the flow of gasoline to the engine.

accelerator pump In the carburetor, a pump (linked to the accelerator) which momentarily enriches the air-fuel mixture when the accelerator is depressed at low speed.

accessories Devices not considered essential to the operation of a vehicle such as the radio, car heater, and electric window lifts.

accumulator A device used in automatic transmissions to cushion the shock of clutch and servo actions.

A.C.I.D. Abbreviation for a four-mode driving-test cycle used to test exhaust emissions or vehicle driveability; the modes are accelerate, cruise, idle, and decelerate.

additive A substance added to gasoline or oil to improve some property of the gasoline or oil.

adjust To bring the parts of a component or system to a specified relationship, dimension, or pressure.

adjustments Necessary or desired changes in clearances, fit, or settings.

adsorb To collect in a very thin layer on the surface of another material.

advance The moving ahead of the ignition spark in relation to piston position; produced by centrifugal or vacuum devices in accordance with engine speed and intake-manifold vacuum.

afterboil Boiling of fuel in the carburetor or coolant in the engine immediately after the engine is stopped.

afterburner On an automobile engine, a type of exhaust manifold that burns any HC and CO remaining in the exhaust gas.

afterrunning The situation in which an engine continues to run after the ignition is turned off. Sometimes referred to as **dieseling.**

aiming screws Horizontal and vertical self-locking adjusting screws, used to aim a headlight and retain it in the proper position.

A.I.R. Abbreviation for *air-injection reactor,* part of a system of exhaust-emission control. See **air-injection system.**

air bags A passive restraint system consisting of balloon-type passenger-safety devices that inflate automatically on vehicle impact.

air bleed An opening into a gasoline passage through which air can pass, or bleed, into the gasoline as it moves through the passage.

air brakes A braking system that uses compressed air to supply the effort required to apply the brakes.

air cleaner A device, mounted on or connected to the carburetor, for filtering dirt and dust out of air being drawn into the engine.

air compressor An engine-driven pump used to supply air under pressure for operating air brakes and air-powered accessories on a vehicle.

air conditioning An accessory system that conditions passenger-compartment air by cleaning, cooling, and drying it.

air-cooled engine An engine that is cooled by the passage of air around the cylinders, and not by the passage of a liquid through water jackets.

air filter A filter that removes dirt and dust particles from air passing through it.

air-fuel mixture The air and fuel traveling to the combustion chamber after being mixed by the carburetor.

air-fuel ratio The proportions of air and fuel (by weight) supplied for combustion.

air gap A small space between parts that are related magnetically, as in an alternator, or electrically, as the electrodes of a spark plug.

Air Guard Name used by American Motors Corporation for the air-injection system of exhaust-emission control.

air horn In the carburetor, a tubular passage on the atmospheric side of the venturi through which the incoming air must pass, and which contains the choke valve.

air impact wrench An air-powered hand-held tool that runs nuts and bolts on and off quickly using a series of sharp, rapid blows created by pressurized air.

air-injection system An exhaust-emission control system; injects air at low pressure into the exhaust manifold or thermal reactor to complete the combustion of unburned hydrocarbons and carbon monoxide in the exhaust gas.

air-inlet valve A movable door, or valve, in the plenum blower assembly; permits the selection of outside air or inside air for both the heating and air-conditioning systems.

air line A hose, pipe, or tube through which air passes.

air nozzle In an air-injection system, the tube through which air is delivered to the exhaust gas.

air-outlet valve A movable door, or valve, in the plenum blower assembly of an air conditioner; directs air either into the heater core or into ductwork that leads to the evaporator.

air pollution Contamination of the air by natural and people-made pollutants.

air pressure Atmospheric pressure; also the pressure produced by an air pump or by compression of air in a cylinder.

air pump Any device for compressing air. In the air-injection system of exhaust-emission control, an engine-driven (belt-driven) pump incorporating a rotor and vanes.

air resistance The drag on a vehicle moving through the air; increases as the square of the speed of the vehicle.

air starting valve A valve which opens to admit compressed air to an engine cylinder to start the engine, and which closes and remains closed after the engine starts.

air suspension Any suspension system that uses contained air for vehicle springing.

air tool Any hand-held tool powered by air.

air wrench See **air impact wrench.**

alcohol A colorless, volatile liquid which, in some forms, can be used as a fuel for racing engines.

alignment The act of lining up, or the state of being in a true line.

alky Performance term for alcohol used as a fuel for racing engines.

allen wrench A type of hexagonal screwdriver that turns a screw having a matching recessed hex head.

alloy A mixture of two or more metals.

alternating current Electric current that flows first in one direction and then in the opposite direction.

alternator In the vehicle electric system, a device that converts mechanical energy into electric energy for charging the battery and operating electrical accessories. Also known as an ac **generator.**

aluminized valve A valve with a thin layer of aluminum sprayed on the valve face, and sometimes on the top of the valve head; the aluminum provides a thin, hard, corrosion-resistant coating.

aluminum cylinder block An engine cylinder block cast from aluminum or aluminum alloy, usually with cast-iron sleeves installed as cylinder bores.

ambient compressor switch In an air conditioner, a switch that energizes the compressor clutch when the outside air temperature is 32°F [0°C] or above, and deenergizes it when the outside air temperature drops below 32°F [0°C].

ambient temperature The temperature of the air surrounding the car.

ammeter A meter for measuring the amount of current (in amperes) flowing through an electric circuit.

amperage The amount of current, in amperes.

ampere A unit of measure for current; one ampere corresponds to a flow of 6.28×10^{18} electrons per second.

anchors Performance term for brakes.

antenna Any device used to pick up radio signals.

antibackfire valve A valve used, in the air-injection system, to prevent backfiring in the exhaust system during deceleration.

antifreeze A chemical, usually ethylene glycol, that is added to the engine coolant to raise the coolant boiling point and lower its freezing point.

antifriction bearing Name given to almost any type of ball, roller, or tapered-roller bearing.

anti-icing system A carburetor system designed to prevent the formation of ice on a surface or in a passage.

antiknock compound An additive put into gasoline to suppress spark knock or detonation—usually a lead compound (which becomes an air pollutant in the engine exhaust, according to some authorities).

antipercolator A vent in the carburetor that opens to release fuel vapors when the throttle is closed; prevents fuel from being pushed out through the fuel nozzle by pressure buildup.

antisiphon system A small passage designed into a carburetor to prevent fuel from siphoning from the float bowl into the engine.

antiskid system A system installed along with the brake system to prevent wheel lockup during braking and, thus, to prevent skidding.

apexes In a Wankel engine using a triangular-type rotor, the peaks or points on the rotor formed by the meeting of two adjoining rotor faces.

arbor press A small hand-operated press used to produce light pressure.

arcing Name given to the spark that jumps the air gap between two electrical conductors; for example, the arcing of the distributor contact points.

armature A part moved by magnetism, or a part moved through a magnetic field to produce current.

asbestos A fiber material that is heat resistant and nonburning; used for brake linings, clutch facings, and gaskets.

aspect ratio The ratio of tire height to width. For example, a G78 tire is 78 percent as high as it is wide. The lower the number, the wider the tire.

aspirator Vacuum pump. Any device that uses a vacuum to draw up gases or small grainy materials; also, the vacuum pump used in catalytic-converter bead replacement.

ATDC Abbreviation for *after top dead center;* any position of the piston between top dead center and bottom dead center, on the downward stroke.

atmosphere The mass of air that surrounds the earth.

atmospheric pollution See **air pollution.**

atmospheric pressure The weight of the atmosphere, per unit area. Atmospheric pressure at sea level is 14.7 psi absolute [1.03 kg/cm^2 (kilograms per square centimeter)]; it decreases as altitude increases.

atom The smallest particle into which an element can be divided.

atomization The spraying of a liquid through a nozzle so that the liquid is broken into a very fine mist.

attrition Wearing down by rubbing or by friction; abrasion.

automatic choke A choke that positions the choke valve automatically in accordance with engine temperature.

automatic level control A suspension system which compensates for variations in load in the rear of the car; positions the rear at a predesigned level regardless of load.

automatic transmission A transmission in which gear ratios are changed automatically, eliminating the necessity of hand-shifting gears.

automotive air pollution Evaporated and unburned fuel and other undesirable by-products of combustion which escape from a motor vehicle into the atmosphere; mainly carbon monoxide (CO), hydrocarbons (HC), nitrogen oxides (NO$_x$), sulfur oxides (SO$_x$), and particulates.

automotive emissions See **automotive air pollution.**

Autronic Eye An electronic device that uses a phototube to select the proper headlight beam automatically.

axis The center line of a rotating part, a symmetrical part, or a circular bore.

axle A crossbar supporting a vehicle and on which one or more wheels turn.

axle ratio The ratio between the rotational speed (rpm) of the drive shaft and that of the driven wheel; gear reduction in the differential, determined by dividing the number of teeth on the ring gear by the number of teeth on the pinion gear.

B

babbitt A metal consisting of tin, antimony, copper, and other metals; used to line bearings.

backfire-suppressor valve An antibackfire valve used in the air-injection system of exhaust-emission control.

backfiring Preexplosion of the air-fuel mixture so that the explosion passes back around the opened intake valve and through the intake manifold and carburetor. Also applied to the loud explosion of overly rich exhaust gas in the exhaust manifold, which exits through the muffler and tail pipe with a loud popping or banging noise.

backlash In gearing, the clearance between the meshing teeth of two gears.

back pressure Pressure in the exhaust manifold of a running engine; affects volumetric efficiency.

bad car Performance term for an extremely fast car.

balanced carburetor Carburetor in which the float bowl is vented into the air horn to compensate for the possible effects of a clogged air filter.

balanced valve A type of hydraulic valve that produces pressure changes proportional to the movement of a mechanical linkage, or to variations in spring pressure.

balancing-coil gauge An indicating device (fuel supply, oil pressure, engine temperature) that contains a pair of coils in the instrument-panel unit.

ball-and-nut steering gear See **recirculating-ball-and-nut steering gear.**

ball-and-trunnion joint A type of universal joint that combines the universal joint and the slip joint in one assembly.

ballast resistor Same as **ignition resistor.**

ball bearing An antifriction bearing with an inner race and an outer race, and one or more rows of balls between them.

ball check valve A valve consisting of a ball and a seat. Fluid can pass in one direction only; flow in the other direction is checked by the ball seating tightly on the seat.

ball joint A flexible joint consisting of a ball within a socket; used in front-suspension systems and valve-train rocker arms.

ball-joint angle The inward tilt of the steering axis from the vertical.

ball-joint suspension A type of front suspension in which the wheel spindle is attached directly to the upper and lower suspension arms through ball joints.

ball-peen hammer A hammer whose head is rounded at one end.

ball stud A stud with a ball-shaped end; commonly used in steering linkages to connect the pitman arm to the linkage, or to connect tie rods.

band In an automatic transmission, a hydraulically controlled brake band installed around a metal clutch drum; used to stop or permit drum rotation.

barrel Term sometimes applied to the cylinders in an engine; used in referring to the number of throttle bores in a carburetor.

base circle The low portion of each cam on a camshaft, which is not part of the lobe.

battery An electrochemical device for storing energy in chemical form so that is can be released as electricity; a group of electric cells connected together.

battery acid The electrolyte used in a battery, a mixture of sulfuric acid and water.

battery cell A battery element that is covered with electrolyte; a cell has a specific gravity of approximately 1.300 and a voltage of approximately 2 volts when fully charged.

battery charge Restoration of chemical energy to a battery by supplying a measured flow of electric current to it over a specified period of time.

battery efficiency The ability of a battery to vary the current it delivers within wide limits, depending on temperature and rate of discharge.

battery element A group of unlike positive and negative plates assembled with separators. There is one element in a cell.

BDC Abbreviation for **bottom dead center.**

bead That part of the tire which is shaped to fit the rim; the bead is made of steel wires, wrapped and reinforced by the plies of the tires.

bearing A part that transmits a load to a support and, in so doing, absorbs the friction of moving parts.

bearing caps In the engine, caps held in place by bolts or nuts which, in turn, hold bearing halves in place.

bearing crush The additional height (over a full half) which is purposely manufactured into each bearing half to ensure complete contact of the bearing back with the housing bore when the engine is assembled.

bearing groove A channel cut in the surface of a bearing to distribute oil.

bearing oil clearance The space purposely provided between a shaft and a bearing, through which lubricating oil can flow.

bearing prelubricator A special tank, attached to an air line; supplies oil to the engine lubricating system at a predetermined and maintained pressure when the engine is not operating.

bearing roll-out tool A special tool, basically a small pin with a thin head; when placed in the crankshaft-journal oil hole, it can be used to roll out or roll in the top half of a main bearing while the crankshaft is still in place.

bearing spin A type of bearing failure in which a lack of lubrication overheats the bearing until it seizes on the shaft, shears its locking lip, and rotates in the housing or block.

bearing spread A purposely manufactured small extra distance across the parting faces of the bearing half, in excess of the actual diameter of the housing bore.

bellows A device, usually metal, that can lengthen or shorten much like an accordian. Some cooling-system thermostats are of the bellows type.

bell-shaped wear Situation in which an opening (such as a brake drum) is worn mostly at one end, so that the opening flares out like a bell.

belt In a tire, a flat strip of material—glass fiber, rayon, or woven steel—which underlies the tread, all around the circumference of the tire.

belted-bias tire A tire in which the plies are laid on the bias, criss-crossing each other, with a circumferential belt on top of them. The rubber tread is vulcanized on top of the belt and plies.

belted-radial tire A tire in which the plies run parallel to each other and perpendicular to the tire bead. Belts running parallel to the tire tread are applied over this radial section.

belt tension The tightness of a drive belt.

bench vise A bench-mounted device having two jaws which may be adjusted to hold an object tightly.

Bendix drive A type of starting-motor drive which screws into mesh with the flywheel teeth as the starting-motor armature begins to turn. It demeshes automatically as the engine speed increases when the engine starts.

bevel gear A gear shaped like the lower part of a cone; used to transmit motion through an angle.

bhp Abbreviation for **brake horsepower.**

bias-belted tire Tire with two extra belts just under the tread; the cords in these belts are set at an angle, or bias, so that the belts hold the tread more firmly to the road.

bias-ply tire A conventionally constructed tire in which the plies are laid on the bias, crisscrossing each other at an angle of about 30° to 40°.

big end The crankpin end of the connecting rod.

bimetal A thermostatic element made up of two metals with different heat expansion rates. Temperature changes produce a bending or distortion of the element.

binders Performance term for brakes.

bite Performance term for tire traction.

bleeding A process by which air is removed from a hydraulic system (brake or power steering) by draining part of the fluid or operating the system to work out the air.

Bloc-Chek A special measuring device that, when inserted in the radiator filler neck of a running engine, can detect the leakage of exhaust gas into the cooling system.

block See **cylinder block.**

blow Performance term for an engine, transmission, or differential failure, as to "blow an engine."

blow-by Leakage of compressed air-fuel mixture and burned gases (from combustion) past the piston rings into the crankcase.

blower Performance term for a supercharger or a two-stroke diesel-engine intake-air compressor. Also, the fan motor in a heater or air-conditioning system.

bmep Abbreviation for *brake mean effective pressure;* the pressure which, acting on the piston, would result in the given brake-horsepower output, if there were no losses due to friction and driving the engine accessories.

body On a vehicle, the assembly of sheet-metal sections, together with windows, doors, seats, and other parts, that provides enclosures for the passengers, engine, etc.

body mounting Putting a car body onto a car chassis. Also, the placing of rubber cushions at strategic points along the chassis to soak up noise and vibration.

body panels Sheets of steel which are fastened together to form the car body.

boiling Conversion from the liquid to the vapor state, taking place throughout the liquid. The conversion is accompanied by bubbling as vapor rises from below the surface.

boiling point The temperature at which a liquid begins to boil.

boil tank A very large tank of boiling parts-cleaning solution, usually used for cleaning cylinder blocks, axle housings, and other large metal parts. Also called a *hot tank*.

bolt A type of fastener having a head on one end and threads on the other; usually used with a nut.

bomb Performance term for a car capable of exceptional performance.

boots Performance term for tires.

borderline lubrication Type of poor lubrication resulting from greasy friction; moving parts are coated with a very thin film of lubricant.

bore An engine cylinder, or any cylindrical hole. Also used to describe the process of enlarging or accurately refinishing a hole, as "to bore an engine cylinder." The bore size is the diameter of the hole.

bore out To increase the engine-cylinder diameter by boring it larger; requires the fitting of oversized pistons. In building a high-performance engine, done to obtain greater engine displacement and power.

boring bar An electric-motor-powered cutting tool used to machine, or bore, an engine cylinder, thereby removing metal and enlarging the cylinder diameter.

bottom dead center The piston position at the lower limit of its travel in the cylinder, such that the cylinder volume is at its maximum.

box Performance term for the transmission.

brake An energy-conversion device used to slow, stop, or hold a vehicle or mechanism. A device which changes the kinetic energy of motion into useless and wasted heat energy.

brake drag A constant, relatively light contact between brake linings and drums when the brakes are not applied. The result is a car that pulls; the brakes destroy themselves by burning up from the generated heat.

brake drum A metal drum mounted on a car wheel to form the outer shell of the brake; the brake shoes press against the drum to slow or stop drum and wheel rotation for braking.

brake-drum glaze Excessively smooth brake-drum surface that lowers friction and, therefore, braking efficiency.

brake fade A reduction, or "fading out," of braking effectiveness; caused by overheating from excessively long and hard brake application, or by water reducing the friction between braking surfaces.

brake feel The reaction of the brake pedal against the driver's foot; tells the driver how heavily the brakes are being applied.

brake fluid A special non-mineral-oil fluid used in the hydraulic braking system to transmit pressure through a closed system of tubing known as the brake lines.

brake grab A sudden increase in braking at a wheel; usually caused by contaminated linings.
brake horsepower Power delivered by the engine and available for driving the vehicle; bhp = torque × rpm/5,252.
brake lines The tubes or hoses connecting the master and wheel cylinders, or calipers, in a hydraulic brake system.
brake lining A high-friction material, usually a form of asbestos, attached to the brake shoe by rivets or a bonding process. The lining takes the wear when the shoe is pressed against the brake drum, or rotor.
brake shoes In drum brakes, arc-shaped metal pieces lined with a high-friction material (the brake lining) which are forced against the revolving drums to produce braking action. In disk brakes, flat metal pieces lined with brake lining which are forced against the rotor face.
brake system A combination of one or more brakes and their operating and control mechanism.
breaker cam See **distributor cam.**
breakerless system An electronic ignition system which does not use mechanical breaker contacts for timing or triggering purposes, but retains the distributor for distribution of the secondary voltage.
breaker points See **contact points.**
breaker-triggered system Any ignition system which utilizes conventional breaker contacts to time and trigger the system; may be a conventional system or an electronic ignition system.
breather On engines without emission-control devices, the opening that allows air to circulate through the crankcase and thus produces crankcase ventilation.
British thermal unit (Btu) A measure of heat quantity. The amount of heat necessary to raise the temperature of 1 lb of liquid water by 1°F.
brush A block of conducting substance, such as carbon, which rests against a rotating ring or commutator to form a continuous electric circuit.
BTDC Abbreviation for *before top dead center;* any position of the piston between bottom dead center and top dead center, on the upward stroke.

bulb An indivisible assembly which contains a source of light; normally used in a lamp.

burr A feather edge of metal left on a part being cut with a file or other cutting tool.

bushing A one-piece sleeve placed in a bore to serve as a bearing surface.

butane A type of liquefied petroleum gas that is liquid below 32°F [0°C] at atmospheric pressure.

butterfly A type of valve used for the choke and throttle valve in a carburetor; a movable flat plate that governs the flow of air into the carburetor.

butyl A type of synthetic rubber used in making tire tubes.

bypass A separate passage which permits a liquid, gas, or electric current to take a path other than that normally used.

bypass valve In an oil filter, a valve that opens when the filter has clogged, to allow oil to reach the engine. See **solenoid bypass valve.**

C

cables Stranded conductors, usually covered with insulating material, used for connections between electrical devices.

cadmium-tip tester A battery tester with two cadmium tips which are inserted into the electrolyte of adjacent battery cells to determine cell voltage.

calibrate To check or correct the initial setting of a test instrument.

caliper A tool that can be set to measure the thickness of a block, the diameter of a shaft, or the bore of a hole (inside caliper). In a disk brake, a housing for pistons and brake shoes, connected to the hydraulic system; holds the brake shoes so that they straddle the disk.

calorie A measure of heat quantity. The amount of heat needed to raise the temperature of 1 g (gram) of water by 1°C.

cam A rotating lobe or eccentric which can be used with a cam follower to change rotary motion to reciprocating motion.

cam angle See **dwell.**

camber The tilt of the top of the wheels from the vertical; when the tilt is outward, the chamber is positive. Also, the angle which a front-wheel spindle makes with the horizontal.

camelback A strip of new rubber tread used to recap a tire.

cam follower See **valve lifter.**

cam-ground piston A piston that is ground slightly oval in shape. It becomes round as it expands with heat.

camshaft The shaft in the engine which has a series of cams for operating the valve mechanisms. It is driven by gears or sprockets and a toothed belt or chain from the crankshaft.

canister A special container, in an evaporative control system, that contains charcoal to trap vapors from the fuel system.

capacitor See **condenser.**

capacitor-discharge ignition system An ignition system which stores its primary energy in a capacitor; available for automobiles, but standard on some outboard engines and motorcycles.

capacity The ability to perform or to hold.

capillary tube A tube with a small inside diameter. In air conditioners, a capillary tube is used to produce a pressure differential between the condenser and the evaporator.

carbon (C) A black deposit left on engine parts such as pistons, rings, and valves by the combustion of fuel, and which inhibits their action.

carbon dioxide (CO_2) A colorless, odorless gas which results from complete combustion; usually considered harmless. The gas absorbed from air by plants in photosynthesis; also used to carbonate beverages.

carbon monoxide (CO) A colorless, odorless, tasteless, poisonous gas which results from incomplete combustion. A pollutant contained in engine exhaust gas.

carbon pile A pile, or stack, of carbon disks enclosed in an insulating tube. When the disks are pressed together, the electrical resistance of the pile decreases.

carburetion The actions that take place in the carburetor: converting liquid fuel to vapor and mixing it with air to form a combustible mixture.

carburetor The device in an engine fuel system which mixes fuel with air and supplies the combustible mixture to the intake manifold.

carburetor heated air A system in which heated air, radiated from the exhaust manifold, is routed to the carburetor for more complete combustion and better engine performance with a leaner air-fuel mixture.

carburetor insulator A spacer, or insulator, used to prevent excess engine heat from reaching the carburetor.

carburetor kickdown Moderate depressing of the accelerator pedal to change the engagement of the choke–fast-idle-speed screw from the high step to a lower step of the cam.

carcinogen or **carcinogenic** A substance or agent that produces or incites cancer.

Cardan universal joint A ball-and-socket type of universal joint.

car lift An air-powered, electrical, or hydraulic piece of shop equipment that can lift an entire vehicle or, in some cases, one end of a vehicle.

case hardening The carburizing method used on low-carbon steel or other alloys to make the case or outer layer of the metal harder than its core.

casing The outer part of the tire assembly, made of fabric or cord to which rubber is vulcanized.

caster Tilting of the steering axis forward or backward to provide directional steering stability. Also, the angle which a front-wheel kingpin makes with the vertical.

catalyst A substance that can speed or slow a chemical reaction between substances, without itself being consumed by the reaction. In the catalytic converter, platinum and palladium are the active catalysts.

catalytic converter A mufflerlike device for use in an exhaust system; it converts harmful exhaust gases into harmless gases by promoting a chemical reaction between a catalyst and the pollutants.

cc Abbreviation for **cubic centimeter.**

CCS Abbreviation for **controlled-combustion system.**

CEC See **combination emission-control system.**

CEC solenoid A two-position electrically operated control used in some TCS systems; either allows or denies distributor vacuum advance, depending on transmission-gear selection. The control-solenoid plunger, when extended, maintains a predetermined throttle opening.

cell Formed by suspending an element of unlike positive and negative plates in electrolyte in a compartment of a battery. The cell produces about 2 volts.

Celsius See **centigrade.**

centigrade A thermometer scale on which water boils at 100° and freezes at 0°. The formula °C = $\frac{5}{9}$(°F − 32) converts readings to centigrade (Celsius) readings.

centimeter (cm) A unit of linear measure in the metric system; equal to approximately 0.39 in.

centrifugal See **centrifugal force.**

centrifugal advance A rotating-weight mechanism in the distributor; advances and retards ignition timing through the centrifugal force resulting from changes in the engine distributor rotational speed.

centrifugal clutch A clutch that uses centrifugal force to apply a higher pressure against the friction disk as the clutch spins faster.

centrifugal filter fan A filter fan mounted on the air-pump drive shaft; used to clean the air entering the air pump.

centrifugal force The force acting on a rotating body and which tends to move it outward and away from the center of rotation. The force increases as rotational speed increases.

ceramic A type of material made from various minerals by baking or firing at high temperatures; can be used as an electrical insulator or as a catalyst substrate in a catalytic converter.

cetane number An indicator of the ignition quality of diesel fuel. A high-cetane fuel ignites more easily (at lower temperature) than a low-cetane fuel.

chain hoist A mechanical device with pulleys and chain; used for lifting heavy objects such as engines.

change of state Transformation of a substance from solid to liquid, from liquid to vapor, or vice versa.

charcoal canister A container filled with activated charcoal, used to trap gasoline vapor from the fuel tank and carburetor while the engine is off.

charge A specific amount of refrigerant or refrigerant oil.

charging rate The amperage flowing from the alternator into the battery.

charging the system The process of adding refrigerant to an air-conditioning system.

chassis The assembly of mechanisms that make up the major operating part of the vehicle; usually assumed to include everything except the car body.

cheater slicks Performance term for slick-tread, soft-compound rear tires.

check To verify that a component, system, or measurement complies with specifications.

check valve A valve that opens to permit the passage of air or fluid in one direction only, or operates to prevent (check) some undesirable action.

chemical instability An undesirable condition caused by the presence of contaminants in a refrigeration system. Refrigerants are stable chemicals, but in contact with contaminants they may break down into harmful chemicals.

chemical reaction The formation of one or more new substances when two or more substances are brought together.

chisel A cutting tool with a specially shaped cutting edge, designed to be driven by a hammer.

choke In the carburetor, a device used when starting a cold engine; it "chokes off" the air flow through the air horn, producing a partial vacuum in the air horn for greater fuel delivery and a richer mixture. Operates automatically on many newer cars.

choke plate In the carburetor, a valve that "chokes off" the air flow through the air horn, producing a partial vacuum in the carburetor for greater fuel delivery and a richer mixture.

Christmas tree Performance term for the set of brightly colored, electronically controlled lights used to start drag races.

chrome-plated ring A piston compression or oil ring with its cylinder-wall face lightly plated with hard chrome.

CID Abbreviation for **cubic inch displacement.**

circuit The complete path of an electric current, including the current source. When the path is continuous, the circuit is closed and current flows. When the path is broken, the circuit is open and no current flows. Also used to refer to fluid paths, as in refrigerant and hydraulic systems.

circuit breaker A protective device that opens an electric circuit to prevent damage when overheated by excess current flow. One type contains a thermostatic blade that warps to open the circuit when the maximum safe current is exceeded.

class Performance term for the category in which a car will race (based on engine displacement).

clearance The space between two moving parts, or between a moving and a stationary part, such as a journal and a bearing. The bearing clearance is filled with lubricating oil when the mechanism is running.

clock Performance term for speedometer. Also means to time a drag-strip run or a lap around a track.

closed-crankcase ventilation system A system in which the crankcase vapors (blow-by gases) are discharged into the engine intake system and pass through to the engine cylinders rather than being discharged into the air.

clutch A coupling which connects and disconnects a shaft from its drive while the drive mechanism is running. In an automobile power train, the device which engages and disengages the transmission from the engine. In an air-conditioning system, the device which engages and disengages the compressor shaft from its continuously rotating drive-belt pulley.

clutch disk See **friction disk.**

clutch fork In the clutch, a Y-shaped member into which the throw-out bearing is assembled.

clutch gear See **clutch shaft.**

clutch housing A metal housing that surrounds the flywheel and clutch assembly.

clutch pedal A pedal in the driver's compartment that operates the clutch.

clutch safety switch See **neutral-start switch.**

clutch shaft The shaft on which the clutch is assembled, with the gear that drives the countershaft in the transmission on one end. On the clutch-gear end, it has external splines that can be used by a synchronizer drum to lock the clutch shaft to the main shaft for direct drive.

clutch solenoid In automotive air conditioners, a solenoid that operates a clutch on the compressor drive pulley. When the clutch is engaged, the compressor is driven and cooling takes place.

CO See **carbon monoxide**.

CO_2 See **carbon dioxide**.

coasting-richer system A system, controlled by a carburetor electromechanical solenoid valve, which provides fuel enrichment while the vehicle is coasting; prevents popping in the exhaust manifold due to the operation of the air-injection system. Used on Chevrolet LUV light trucks.

coated ring A piston ring with its cylinder-wall face coated with ferrous oxide, soft phosphate, or tin. This thin coating helps new rings seat by retaining oil and reducing scuffing during break-in.

coefficient of expansion The proportional increase in a dimension of an object per degree of temperature rise.

coil In an automobile ignition system, a transformer used to step up the battery voltage (by induction) to the high voltage required to fire the spark plugs.

coil spring A spring made of an elastic metal such as steel, formed into a wire and wound into a coil.

coil-spring clutch A clutch using coil springs to hold the pressure plate against the friction disk.

cold The absence of heat. An object is considered cold to the touch if its temperature is less than body temperature (98.6°F [37°C]).

cold-cranking rate A battery rating; the minimum amperage maintained by a battery for 30 seconds with a minimum voltage of 1.2 volts per cell, checked at a battery temperature of 0°F [−17.8°C] and at −20°F [−28.9°C].

cold-patching A method of repairing a punctured tire or tube by gluing a thin rubber patch over the hole.

cold rate A battery rating; the number of minutes a battery will deliver 300 A at 0°F [−17.8°C] before the cell voltage drops below 1.0 volt.

cold-start test A prescribed federal test procedure for measuring emissions from a vehicle before the engine is warmed up. The test is made on the vehicle after a 12-hour cold-soak period at 68 to 78°F [20 to 25°C].

cold welding Repairing a crack in metal by drilling a hole through the crack, threading the hole, and screwing in a section of threaded rod to form a seal.

collapsible steering column An energy-absorbing steering column designed to collapse if the driver is thrown into it by a severe collision.

combination emission control system An exhaust-emission control system, used on some General Motors cars, which combines a transmission-controlled spark system and a deceleration throttle-positioning device.

combustion Burning; fire produced by the proper combination of fuel, heat, and oxygen. In the engine, the rapid burning of the air-fuel mixture in the combustion chamber.

combustion chamber The space between the top of the piston and the cylinder head, in which the air-fuel mixture is burned.

commutation In a dc generator, the effect produced by the commutator and brushes, in which the alternating current developed in the armature windings is changed to direct current.

commutator A series of copper bars at one end of a generator or starting-motor armature, electrically insulated from the armature shaft and insulated from each other by mica. The brushes rub against the bars of the commutator, which form a rotating connector between the armature windings and brushes.

compensating port A small hole, in each section of a master cylinder, which is closed by piston movement so that fluid is trapped ahead of the piston to apply the brakes. When the pedal is released, the piston uncovers the compensating port, allowing the trapped fluid to return to the fluid reservoir.

compound vortex-controlled combustion engine A type

of stratified-charge engine built by Honda. See **Honda system.**

compression Reducing the volume of a gas by squeezing it into a smaller space. Increasing the pressure reduces the volume and increases the density and temperature of the gas.

compression ignition The ignition of fuel solely by the heat generated when air is compressed in the cylinder; the method of ignition in a diesel engine.

compression pressure The pressure in the combustion chamber at the end of the compression stroke.

compression ratio The volume of the cylinder and combustion chamber when the piston is at BDC, divided by the volume when the piston is at TDC.

compression ring The upper ring or rings on a piston, designed to hold the compression in the combustion chamber and prevent blow-by.

compression stroke The piston movement from BDC to TDC immediately following the intake stroke, during which both the intake and exhaust valves are closed while the air-fuel mixture in the cylinder is compressed.

compression tester An instrument for testing the amount of pressure, or compression, developed in an engine cylinder during cranking.

compressor The component of an air-conditioning system that compresses refrigerant vapor to increase its pressure and temperature.

condensate Water that is removed from air. It forms on the exterior surface of the air-conditioner evaporator.

condensation A change of state during which a gas turns to liquid, usually because of temperature or pressure changes. Also, moisture from the air, deposited on a cool surface.

condenser In the ignition system, a device that is also called a *capacitor;* connected across the contact points to reduce arcing by providing a storage place for electricity (electrons) as the contact points open. In an air-conditioning system, the radiatorlike heat exchanger in which refrigerant vapor loses heat and returns to the liquid state.

conditioned air Cool, dry, clean air.

conduction The transfer of heat between the closely packed molecules of a substance or between two substances that are touching.

conductor Any material or substance that allows current or heat to flow easily.

connecting rod In the engine, the rod that connects the crank on the crankshaft with the piston. Sometimes called a *con rod.*

connecting-rod bearing See **rod bearing.**

connecting-rod cap The part of the connecting-rod assembly that attaches the rod to the crankpin.

constant-current charging A battery-charging method in which an unchanging amount of current is made to flow into the battery.

constant-velocity joint Two closely coupled universal joints arranged so their acceleration-deceleration effects cancel each other out. This results in an output drive-shaft speed that is always identical with the input drive-shaft speed, regardless of the angle of drive.

constant-voltage charging A charging method in which a constant voltage is applied to the battery. The charging current decreases as the battery approaches the charged condition.

contact points In the conventional ignition system, the stationary and the movable points in the distributor which open and close the ignition primary circuit.

contaminants Anything other than refrigerant and refrigerant oil in a refrigeration system; includes rust, dirt, moisture, and air.

control arm A part of the suspension system designed to control wheel movement precisely.

controlled-combustion system An exhaust-emission control system used by General Motors; regulates engine combustion efficiency through special settings of the carburetor, distributor, and vacuum advance, by heating the carburetor intake air, and with a higher engine operating temperature. Also known as the *engine modification system,* and used by other manufacturers under other names.

convection The transfer of heat by motion of the heated

23

material. Moving currents of heated liquid or gas are called *convection currents*.

coolant The liquid mixture of about 50 percent antifreeze and 50 percent water used to carry heat out of the engine.

cooling system The system that removes heat from the engine by the forced circulation of coolant, and thereby prevents engine overheating. It includes the water jackets, water pump, radiator, and thermostat.

core In a radiator, a number of coolant passages surrounded by fins through which air flows to carry away heat.

cornering wear A type of tire-tread wear caused by turning at excessive speeds.

corrosion Chemical action, usually by an acid, that eats away (decomposes) a metal.

cotter pin A type of fastener, made from soft steel in the form of a split pin, that can be inserted in a drilled hole. The split ends are spread to lock the pin in position.

counterbored ring A piston ring, used as a compression ring, which has a counterbore on its inside diameter to promote cylinder sealing.

countershaft The shaft, in the transmission, which is driven by the clutch gear; gears on the countershaft drive gears on the main shaft when the latter are shifted into gear.

counterweight A weight mounted on the crankshaft opposite each crankpin; reduces the vibration and bearing loads due to the inertia of moving parts.

crane In the automotive shop, a portable hoist used for removing and installing engines.

crank Slang for **crankshaft**.

crankcase The lower part of the engine in which the crankshaft rotates; includes the lower section of the cylinder block and the oil pan.

crankcase breather The opening or tube that allows air to enter and leave the crankcase and thus permit crankcase ventilation.

crankcase dilution Dilution of the lubricating oil in the oil pan; caused by liquid gasoline condensing from the blow-by in a cold engine and seeping down the cylinder walls.

crankcase emissions Pollutants emitted into the atmosphere from any portion of the engine-crankcase ventilation or lubrication system.

crankcase ventilation The circulation of air through the crankcase of a running engine to remove water, blow-by, and other vapors; prevents oil dilution, contamination, sludge formation, and pressure buildup.

cranking motor See **starting motor.**

crankpin The part of a crankshaft to which a connecting rod is attached.

crankpin ridging A type of crankpin failure, typified by deep ridges worn into the crankpin bearing surfaces.

crankshaft The main rotating member or shaft of the engine, with cranks to which the connecting rods are attached; converts up and down (reciprocating) motion into circular (rotary) motion.

crankshaft gauge A special type of micrometer which can measure crankshaft wear while the crankshaft is in the cylinder block.

crankshaft gear A gear, or sprocket, mounted on the front of the crankshaft; used to drive the camshaft gear, chain, or toothed belt.

crank throw One crankpin with its two webs.

crank web The part of the crankshaft that lies between a crankpin and a main bearing of the crankshaft.

cross-firing Jumping of a high-voltage surge in the ignition secondary circuit to the wrong high-voltage lead, so that the wrong spark plug fires. Usually caused by improper routing of the spark-plug wires, faulty insulation, or a defective distributor cap or rotor.

cross-flow radiator A radiator in which the coolant flows horizontally from the input tank on one side of the radiator, through the individual coolant passages, to the output tank on the opposite side of the radiator.

crowd An acceleration that maintains a constant manifold-vacuum reading. This requires a progressive opening of the throttle as the vehicle speed increases.

CRS Abbreviation for **coasting-richer system.**

cubic centimeter (cucm or cc) A unit of volume in the metric system; equal to approximately 0.061 in^3.

cubic inch displacement The cylinder volume swept out by the pistons of an engine as they move from BDC to TDC, measured in cubic inches.

curb weight The weight of an empty vehicle without payload or driver but including fuel, coolant, oil, and all items of standard equipment.

current A flow of electrons, measured in amperes.

cut out In a running engine, to miss momentarily but not stall.

cutout relay A device in the charging circuit between the generator and battery; closes when the generator charges the battery, and opens when the generator stops.

CVCC See **compound vortex-controlled combustion;** also, **Honda system.**

cycle Any series of events which repeat continuously. In the engine, the four (or two) piston strokes that together produce the power.

cycling-clutch system An air conditioner in which the conditioned-air temperature is controlled by starting and stopping the compressor.

cylinder A circular tubelike opening in an engine cylinder block or casting in which a piston moves up and down.

cylinder block The basic framework of the engine, in and on which the other engine parts are attached. It includes the engine cylinders and the upper part of the crankcase.

cylinder compression tester See **compression tester.**

cylinder head The part of the engine that covers and encloses the cylinders. It contains cooling fins or water jackets and, on I-head engines, the valves.

cylinder hone An expandable rotating tool with abrasive fingers turned by an electric motor; used to clean and smooth the inside surface of a cylinder.

cylinder leakage tester A testing device that forces compressed air into the cylinder through the spark-plug hole, when the valves are closed and the piston is at TDC on the compression stroke. The percentage of compressed air that leaks out is measured, and the source of the leak accurately pinpoints the defective part.

cylinder liner See **cylinder sleeve.**

cylinder sleeve A replaceable sleeve, or liner, set into the cylinder block to form the cylinder bore.

D

dashpot A device on the carburetor that prevents the throttle valve from closing too suddenly.

DC or **dc** Abbreviation for **direct current.**

dead axle An axle that simply supports weight and attached parts, but does not turn or deliver power to a wheel or other rotating member.

deceleration A decrease in velocity or speed. Also, allowing the car or engine to coast to idle speed from a higher speed with the accelerator at or near the idle position.

deceleration valve A device used in conjunction with the dual-diaphragm vacuum-advance unit to advance the timing under deceleration conditions.

deflection rate For a spring, the number of pounds required to compress the spring exactly 1 in [25.4 mm].

defroster The part of the car heater system designed to melt frost or ice on the inside or outside of the windshield; includes the required ductwork.

degree Part of a circle. One degree is $\frac{1}{360}$ of a complete circle.

dehumidify To remove water vapor from the air. In an air conditioner, the air is dehumidified as it passes through the evaporator, since water condenses from the air onto the cool evaporator coils.

dehydrator-filter In an air conditioner, a filtering device in the refrigerant line between the condenser and evaporator; removes dirt and moisture from the liquid refrigerant.

Delco Eye A type of battery vent cap that shows a low electrolyte level in the cell without being removed from the battery.

desiccant A drying agent. In a refrigeration system, desiccant is placed in the receiver-dehydrator to remove moisture from the system.

detent A small depression in a shaft, rail, or rod into which a pawl or ball drops when the shaft, rail, or rod is moved; this provides a locking effect.

detergent A chemical added to engine oil; helps keep internal parts of the engine clean by preventing the accumulation of deposits.

detonation Commonly referred to as spark knock or **ping.** In the combustion chamber, an uncontrolled second explosion (after the spark occurs at the spark plug) with spontaneous combustion of the remaining compressed air-fuel mixture, resulting in a pinging noise.

device A mechanism, tool, or other piece of equipment designed to serve a special purpose or perform a special function.

diagnosis A procedure followed in locating the cause of a malfunction.

dial indicator A gauge that has a dial face and a needle to register movement; used to measure variations in dimensions, and movements too small to be measured conveniently by other means.

diaphragm A thin dividing sheet or partition which separates an area into compartments; used in fuel pumps, modulator valves, vacuum-advance units, and other control devices.

diaphragm spring A spring shaped like a disk with tapering fingers pointed inward, or like a wavy disk (crown type).

diaphragm-spring clutch A clutch in which a diaphragm spring, rather than a coil spring, applies pressure against the friction disk.

dice Performance term meaning to compete against another car.

dichlorodifluoromethane Refrigerant-12, whose chemical formula is CCl_2F_2.

die A tool for cutting threads on a rod.

die out To stall or stop running, as an engine.

diesel cycle An engine operating cycle in which air is compressed, and fuel oil is injected into the compressed air at the end of the compression stroke. The heat produced by the compression ignites the fuel oil, eliminating the need for spark plugs or a separate ignition system.

diesel engine An engine operating on the diesel cycle and burning oil instead of gasoline.

dieseling A condition in which an automobile engine continues to run after the ignition is off. Caused by carbon deposits or hot spots in the combustion chamber glowing sufficiently to furnish heat for combustion.

differential A gear assembly between axles that permits one wheel to turn at a different speed than the other, while transmitting power from the drive shaft to the wheel axles.

differential case The metal unit that encases the differential pinions and side gears, and to which the ring gear is attached.

differential side gears The gears inside the differential case which are internally splined to the axle shafts, and which are driven by the differential pinion gears.

dig out Performance term meaning to accelerate rapidly from standing start.

dimmer switch A two-position switch, usually mounted on the car floor; operated by the driver to select the high or low headlight beam.

diode A solid-state electronic device which allows the passage of an electric current in one direction only. Used in the alternator to convert alternating current to direct current for charging the battery.

dipstick See **oil-level indicator.**

direct-bonded bearing A bearing formed by pouring liquid babbitt (bearing metal) directly into the bearing housing, and machining the cooled metal to the desired bearing diameter.

direct current Electric current that flows in one direction only.

directional signal A device on the car that flashes lights to indicate the direction in which the driver intends to turn.

disassemble To take apart.

disc brake Brake in which brake shoes, on a vise-like caliper, grip a revolving disk to stop it.

discharge To depressurize; to crack a valve to allow refrigerant to escape from an air conditioner; to bleed.

discharge air Conditioned air leaving the refrigeration unit of an air conditioner.

discharge line In an air conditioner, the tube that connects the compressor outlet and the condenser inlet. High-pressure refrigerant vapor flows through the line.

discharge pressure In an air conditioner, the pressure of refrigerant being discharged from the compressor; also called the *high pressure*.

discharge side In an air conditioner, the portion of the refrigerant system under high pressure; extends from the compressor outlet to the thermostatic expansion valve.

disk In a disk brake, the rotor, or revolving piece of metal, against which shoes are pressed to provide braking action.

disk runout The amount by which a brake disk wobbles during rotation.

dispersant A chemical added to oil to prevent dirt and impurities from clinging together in lumps that could clog the engine lubricating system.

displacement In an engine, the total volume of air-fuel mixture an engine is theoretically capable of drawing into all cylinders during one operating cycle. Also, the volume swept out by the piston in moving from one end of a stroke to the other.

distributor Any device that distributes. In the ignition system, the rotary switch that directs high-voltage surges to the engine cylinders in the proper sequence. See **ignition distributor.**

distributor cam The cam on the top end of the distributor shaft which rotates to open and close the contact points.

distributor advance See **centrifugal advance, ignition advance,** and **vacuum advance.**

distributorless ignition An electronic ignition system which does not utilize breaker contacts to time or trigger the system, and does not utilize a distributor for distribution of the secondary voltage. One type is crankshaft-triggered.

distributor plate The plate in the ignition distributor that is fastened to the distributor housing and does not move.

distributor timing See **ignition timing.**

distributor vacuum-advance control valve See **deceleration valve.**

diverter valve In the air-injection system of exhaust-emission control, a valve that diverts air-pump output into the air cleaner or the atmosphere during deceleration; prevents backfiring and popping in the exhaust system.

DOHC See **double-overhead-camshaft engine.**

dolly blocks Blocks of metal, variously shaped and contoured, used to straighten body panels and fenders. The dolly block is held on one side of the panel while the other side is struck with a special hammer.

double-Cardan joint A near-constant-velocity universal joint which consists of two Cardan universal joints connected by a coupling yoke.

double-overhead-camshaft (DOHC) engine An engine with two camshafts in each cylinder head to actuate the valves; one camshaft operates the intake valves, and the other operates the exhaust valves.

double-reduction differential A differential containing an extra set of gears to provide additional gear reduction.

dowel A metal pin attached to one object which, when inserted into a hole in another object, ensures proper alignment.

downdraft carburetor A carburetor in which the air horn is so arranged that the air passes down through it on its way to the intake manifold.

down-flow radiator A radiator in which the coolant enters the radiator at the top, and loses heat as it flows down through passages to the bottom of the radiator.

drafting Performance term for the racing technique of following closely behind another car to reduce wind resistance and save fuel.

drag Performance term for a $\frac{1}{4}$-mile race from a standing start, against time or another car.

driveability The general operation of an automobile, usually rated from good to poor; based on characteristics of concern to the average driver, such as smoothness of idle, even acceleration, ease of starting, quick warmup, and tendency to overheat at idle.

drive line The driving connection between the transmission and the differential; made up of one or more drive shafts.

driven disk The friction disk in a clutch.

drive pinion A rotating shaft that transmits torque to another gear; used in the differential. Also called the *clutch shaft,* in the transmission.

drive shaft An assembly of one or two universal joints and slip joints connected to a heavy metal tube; used to transmit power from the transmission to the differential. Also called the *propeller shaft.*

drop-center wheel The conventional passenger-car wheel, which has a space (drop) in the center for one bead to fit into while the other bead is being lifted over the rim flange.

drop light A portable light, with a long electric cord, used in the shop to illuminate the immediate work area.

drum brakes Brakes in which curved brake shoes press against the inner circumference of a metal drum to produce the braking action.

drum lathe Special lathe for turning brake drums; some can be used to resurface disk-brake rotors.

dry-charged battery A new battery that has been charged, and then stored with the electrolyte removed. Electrolyte must be added to activate the battery at the time of sale.

dry-disk clutch A clutch in which the friction faces of the friction disk are dry, as opposed to a wet-disk clutch which runs submerged in oil. The conventional type of automobile clutch.

dry friction The friction between two dry solids.

drying agent See **desiccant.**

dry sump A type of engine-lubrication system in which the oil supply is carried in a separate oil tank instead of in the conventional oil pan.

dual-area diaphragm An automatic-transmission shift-control diaphragm which has sources of vacuum from the intake manifold and EGR port.

dual-brake system A brake system consisting of two separate hydraulic systems; usually, one operates the front brakes, and the other operates the rear brakes.

dual carburetors Two carburetors mounted on one engine.

dual-diaphragm advance A vacuum-advance mechanism with two diaphragms; attaches to the engine distributor to control spark timing. One diaphragm provides normal ignition timing advance for starting and acceleration; the other diaphragm retards the spark during idling and part-throttle operation.

dual-point system A system that controls spark timing by electromechanical selection of separate advance and retard distributor points; used on Chevrolet LUV light trucks. Sometimes used to refer to any ignition system which has two sets of contact points in the distributor.

dual quad Performance term for a carburetion setup that uses two four-barrel carburetors.

duct A tube or channel used to convey air or liquid from one point to another. In emission systems, a tube on an air cleaner that has a vacuum motor mounted on it to help regulate the temperature of the carburetor intake air.

durability The quality of being useful for a long period of time and service. Used to indicate the useful life of a catalyst or emission-control system.

dwell The number of degrees the distributor shaft or cam rotates while the distributor points are closed.

dwell meter A precision electrical instrument used to measure the cam or dwell or number of degrees the distributor points are closed while the engine is running.

Dyer drive A type of starting-motor drive, used on heavy-duty applications, which provides mechanical meshing of the drive pinion (as in an overrunning clutch) and automatic demeshing (as in a Bendix drive).

dynamic balance The balance of an object when it is in motion (for example, the dynamic balance of a rotating wheel).

dynamometer A device for measuring the power output, or brake horsepower, of an engine. An *engine* dynamometer measures the power output at the flywheel; a *chassis* dynamometer measures the power output at the drive wheels.

E

eccentric A disk or offset section (of a shaft, for example) used to convert rotary to reciprocating motion. Sometimes called a **cam.**

eccentric shaft In a Wankel rotary engine, the main shaft, or crankshaft.

efficiency The ratio between the power of an effect and the power expended to produce the effect; the ratio between an actual result and the theoretically possible result.

EGR system Abbreviation for **exhaust-gas recirculation system.**

electric-assist choke A choke in which a small electric heating element warms the choke spring, causing it to release more quickly. This reduces exhaust emissions during the startup of a cold engine.

electric brakes A braking system with an armature-electromagnet combination at each wheel; when the electromagnet is energized, the magnetic attraction between the armature and electromagnet causes the brake shoes to move against the brake drum.

electric current A movement of electrons through a conductor such as a copper wire; measured in amperes.

electric system In the automobile, the system that electrically cranks the engine for starting; furnishes high-voltage sparks to the engine cylinders to fire the compressed air-fuel charges; lights the lights; and powers the heater motor, radio, and other accessories. Consists, in part, of the starting motor, wiring, battery, alternator, regulator, ignition distributor, and ignition coil.

electrode In a spark plug, the spark jumps between two electrodes. The wire passing through the insulator is the *center* electrode. The small piece of metal welded to the spark-plug shell (and to which the spark jumps) is the *side,* or *ground,* electrode.

electrolyte The mixture of sulfuric acid and water used in lead-acid storage batteries. The acid enters into chemical reaction with active material in the plates to produce voltage and current.

electromagnet A coil of wire (usually around an iron core) which produces magnetism as electric current passes through it.

electromagnetic induction The characteristic of a magnetic field that causes an electric current to be created in a conductor if it passes through the field, or if the field builds and collapses around the conductor.

electron A negatively charged particle that circles the nucleus of an atom. The movement of electrons is an electric current.

electronic fuel-injection system A system that injects gasoline into a spark-ignition engine, and that includes an electronic control to time and meter the fuel flow.

electronic ignition system A transistorized ignition system which does not have mechanical contact points in the distributor, but uses the distributor for distributing the secondary voltage to the spark plugs. Also called a *solid-state* ignition system.

electronic spark control A system that controls the vacuum to the distributor, preventing vacuum advance below a selected vehicle speed; generally used by Ford on cars with automatic transmission.

element A substance that cannot be further divided into a simpler substance. In a battery, the group of unlike positive and negative plates, separated by insulators, that make up each cell.

eliminator Performance term for the fastest drag car in its class.

emission control Any device or modification added onto or designed into a motor vehicle for the purpose of reducing air-polluting emissions.

emission standards Allowable automobile emission levels, set by local, state, and federal legislation.

emitter An engine with considerable exhaust emissions; sometimes preceded by the word "high" or "low" to indicate the degree of emission.

end play As applied to a crankshaft, the distance that the crankshaft can move forward and back in the cylinder block.

energy The capacity or ability to do work. Usually measured in work units of pound-feet [kilogram-meters], but also expressed in heat-energy units (Btus [joules]).

engine A machine that converts heat energy into mechanical energy. A device that burns fuel to produce mechanical power; sometimes referred to as a **power plant.**

engine fan See **fan**.

engine tune-up A procedure for inspecting, testing, and adjusting an engine, and replacing any worn parts, to restore the engine to its best performance.

engine vacuum gauge See **vacuum gauge**.

Environmental Protection Agency The independent agency of the United States government that sets standards and coordinates activities related to automotive emissions and the environment.

EPA Abbreviation for **Environmental Protection Agency**.

epoxy A plastic compound that can be used to repair some types of cracks in metal.

equalizer line In an air conditioner, a line or connection used specifically for obtaining required operation from certain types of control valves. Very little (if any) refrigerant flows through this line.

ESC Abbreviation for **electronic spark control**.

ET Abbreviation for the performance term *elapsed time,* in drag-racing events.

ethyl See **tetraethyl lead**.

ethylene glycol Chemical name of a widely used type of permanent antifreeze.

evacuate To use a vacuum pump to pump any air and moisture out of an air-conditioner refrigerant system; required whenever any component in the refrigerant system has been removed and replaced.

evaporation The transforming of a liquid to the gaseous state.

evaporation control system A system which prevents the escape of gasoline vapors from the fuel tank or carburetor to the atmosphere while the engine is off. The vapors are stored in a charcoal canister or in the engine crankcase until the engine is started.

evaporator The heat exchanger in an air conditioner in which refrigerant changes from a liquid to a gas (evaporates), taking heat from the surrounding air as it does so.

exhaust emissions Pollutants emitted into the atmosphere through any opening downstream of the exhaust ports of an engine.

exhaust gas The burned and unburned gases that remain (from the air-fuel mixture) after combustion.

exhaust-gas analyzer A device for sensing the amounts of air pollutants in the exhaust gas of a motor vehicle. The analyzers used in automotive shops check HC and CO; those used in testing laboratories can also check NO_x.

exhaust-gas recirculation system An NO_x control system that recycles a small part of the inert exhaust gas back through the intake manifold at all throttle positions except idle and wide open, to lower the combustion temperature.

exhaust manifold A device with several passages through which exhaust gases leave the engine combustion chambers and enter the exhaust piping system.

exhaust pipe The pipe connecting the exhaust manifold with the muffler.

exhaust stroke The piston stroke (from BDC to TDC) immediately following the power stroke, during which the exhaust valve opens so that the exhaust gases can escape from the cylinder to the exhaust manifold.

exhaust system The system through which exhaust gases leave the vehicle. Consists of the exhaust manifold, exhaust pipe, muffler, tail pipe, and resonator (if used).

exhaust valve The valve that opens during the exhaust stroke to allow burned gases to flow from the cylinder to the exhaust manifold.

expansion control See **cam-ground piston.**

expansion plug A slightly dished plug that is used to seal core passages in the cylinder block and cylinder head. When driven into place, it is flattened and expanded to fit tightly.

expansion tank A tank at the top of an automobile radiator which provides room for heated coolant to expand and to give off any air that may be trapped in the coolant. Also, a similar device used in some fuel tanks to prevent fuel from spilling out of the tank through expansion.

expansion valve A metering valve or device located between the condenser and evaporator in an air conditioner; controls the amount of refrigerant sprayed into the evaporator.

extreme-pressure lubricant A special lubricant for use in hypoid-gear differentials; needed because of the heavy wiping loads imposed on the gear teeth.

F

fan The bladed device on the front of the engine that rotates to draw cooling air through the radiator, or around the engine cylinders; an air blower such as the heater fan and the A/C blower.

fast flushing A method of cleaning the cooling system; uses a special machine to circulate the cleaning solution.

fast-idle cam A mechanism on the carburetor, connected to the automatic choke, that holds the throttle valve slightly open when the engine is cold; causes the engine to idle at a higher rpm as long as the choke is applied.

fatigue failure A type of metal failure resulting from repeated stress which finally alters the character of the metal so that it cracks. In engine bearings, frequently caused by excessive idling, or slow engine idling speed.

feeler gauge Strips of metal with accurate, known thicknesses; used to measure clearances.

F-head engine An engine with some of the valves in the cylinder head and some in the cylinder block, giving the engine an F-shaped appearance.

field coil A coil, or winding, in a generator or starting motor which produces a magnetic field as current passes through it.

field-frame assembly The round, soft-iron frame in a generator or motor into which the field coils are assembled.

field relay A relay that is part of some alternator charging systems; connects the alternator field to the battery when the engine runs, and disconnects it when the engine stops.

field winding See **field coil**.

fifth wheel A coupling device mounted on a tractor and used to connect a semitrailer. It acts as a hinge to allow the tractor and semitrailer to change direction independently.

file A cutting tool with a large number of cutting edges arranged along a surface.

filter A device through which air, gases, or liquids are passed to remove impurities.

fins On a radiator or heat exchanger, thin metal projections over which cooling air flows to remove heat from hot liquid flowing through internal passages. On an air-cooled engine, thin metal projections on the cylinder and head which greatly increase the area of the heat-radiating surfaces and help cool the engine.

firing line The high-voltage vertical spike, or line, that appears on the oscilloscope pattern of the ignition-system secondary circuit. The firing line shows when the spark plug begins to fire, and the voltage required to fire it.

firing order The order in which the engine cylinders fire, or deliver their power strokes, beginning with No. 1 cylinder.

fixed-caliper disk brake Disk brake using a caliper which is fixed in position and cannot move; the caliper has four pistons, two on each side of the disk.

flasher An automatic-reset circuit breaker used in the directional-signal and emergency-signal circuits.

flat-head engine See **L-head engine.**

flat out Performance term meaning at top speed.

flat rate Method of paying mechanics and technicians, by use of a manual which indicates the time normally required to do each service job.

flat spot Lack of normal acceleration or response to throttle opening; implies no loss of power but also no increase in power.

float bowl In a carburetor, the reservoir from which gasoline is metered into the passing air.

floating-caliper disk brake Disk brake using a caliper mounted through rubber bushings which permit the caliper to float, or move, when the brakes are applied; there is one large piston in each caliper.

float level The float position at which the needle valve closes the fuel inlet to the carburetor, to prevent further delivery of fuel.

float system In the carburetor, the system that controls the entry of fuel and the fuel level in the float bowl.

flooded Term used to indicate that the engine cylinders

received "raw" or liquid gasoline, or an air-fuel mixture too rich to burn.

floor jack A small, portable, hydraulically operated lifting device used to raise part of a vehicle from the floor for repair.

fluid Any liquid or gas.

fluid coupling A device in the power train consisting of two rotating members; transmits power from the engine, through a fluid, to the remainder of the power train.

flush In an air conditioner, to wash out the refrigerant passages with Refrigerant-12, to remove contaminants. In a brake system, to wash out the hydraulic system and the master and wheel cylinders, or calipers, with clean brake fluid to remove dirt or impurities that have gotten into the system.

flywheel A heavy metal wheel that is attached to the crankshaft and rotates with it; helps smooth out the power surges from the engine power strokes; also serves as part of the clutch and engine cranking system.

flywheel ring gear A gear, fitted around the flywheel, that is engaged by teeth on the starting-motor drive to crank the engine.

fog lamp A light which may be mounted to provide illumination forward of the vehicle; used with the lower-beam headlights to provide road illumination under conditions of rain, snow, dust, or fog.

Folo-Thru drive A type of inertia starting-motor drive; similar to the Bendix drive, except that it has a locking pin to hold the pinion in mesh until the engine starts.

force Any push or pull exerted on an object; measured in units of weight, such as pounds, ounces, kilograms, or grams.

four-barrel carburetor A carburetor with four throttle valves. In effect, two two-barrel carburetors in a single assembly.

four-cycle See **four-stroke cycle.**

four on the floor Slang for a four-speed transmission with the shift lever mounted on the floor of the driving compartment, frequently as part of a center console.

four-speed In manual transmissions, having four forward gears.

four-stroke cycle The four piston strokes—intake, compression, power, and exhaust—that make up the complete cycle of events in the four-stroke-cycle engine. Also called *four-cycle* and *four-stroke.*

421 tester A tester for batteries with a one-piece cover; applies timed discharge-charge cycles to the battery to determine its condition.

four-wheel drive On a vehicle, driving axles at both front and rear, so that all four wheels can be driven.

frame The assembly of metal structural parts and channel sections that supports the car engine and body and is supported by the wheels.

frame gauges Gauges that may be hung from the car frame to check its alignment.

Freon-12 Refrigerant used in automobile air conditioners. Also known as **Refrigerant-12** and *R-12.*

friction The resistance to motion between two bodies in contact with each other.

friction bearing Bearing in which there is sliding contact between the moving surfaces. Sleeve bearings, such as those used in connecting rods, are friction bearings.

friction disk In the clutch, a flat disk, faced on both sides with friction material and splined to the clutch shaft. It is positioned between the clutch pressure plate and the engine flywheel. Also called the *clutch disk* or *driven disk.*

friction horsepower The power used up by an engine in overcoming its own internal friction; usually increases as engine speed increases.

front-end drive A vehicle having its drive wheels located on the front axle.

front-end geometry The angular relationship between the front wheels, wheel-attaching parts, and car frame. Includes camber, caster, steering-axis inclination, toe-in, and toe-out on turns.

fuel Any combustible substance. In an automobile engine, the fuel (gasoline) is burned, and the heat of combustion expands the resulting gases, which force the piston downward and rotate the crankshaft.

fuel decel valve A device which supplies additional air-fuel mixture to the intake manifold during deceleration to control exhaust-gas hydrocarbon emissions.

fuel filter A device located in the fuel line, ahead of the float bowl; removes dirt and other contaminants from fuel passing through.

fuel gauge A gauge that indicates the amount of fuel in the fuel tank.

fuel-injection system A system which delivers fuel under pressure into the combustion chamber, or into the air flow just as it enters each individual cylinder. Replaces the conventional carburetor.

fuel line The pipe or tubes through which fuel flows from the fuel tank to the carburetor.

fuel nozzle The tube in the carburetor through which gasoline feeds from the float bowl into the passing air. In a fuel-injection system, the tube that delivers the fuel into the compressed air or the passing air stream.

fuel pump The electrical or mechanical device in the fuel system which forces fuel from the fuel tank to the carburetor.

fuel system In an automobile, the system that delivers the combustible mixture of vaporized fuel and air to the engine cylinders. Consists of the fuel tank and lines, gauge, fuel pump, carburetor, and intake manifold.

fuel tank The storage tank for fuel on the vehicle.

fuel-vapor recovery system See **vapor recovery system.**

full bore Performance term for wide-open throttle.

full coil suspension A vehicle suspension system in which each of the four wheels has its own coil spring.

full-floating piston pin A piston pin that is free to turn in the piston bosses and in the connecting rod.

full-floating rear axle An axle which only transmits driving forces to the rear wheels; the weight of the vehicle (including payload) is supported by the axle housing.

full-flow filter Type of oil filter designed so that all the oil from the oil pump flows through it.

full throttle Wide-open throttle position, with the accelerator pressed all the way down to the floorboard.

fuse A device designed to open an electric circuit when the current is excessive, to protect equipment in the circuit. An open, or "blown," fuse must be replaced after the circuit problem is corrected.

fuse block A boxlike unit that holds the fuses for the various electric circuits in an automobile.

fusible link A type of fuse in which a special wire melts to open the circuit when the current is excessive. An open, or "blown," fusible link must be replaced after the circuit problem is corrected.

fusion Melting; conversion from the solid to the liquid state.

G

gallery A passageway inside a wall or casting. The main oil gallery within the block supplies lubrication to all parts of the engine.

gap The air space between two electrodes, as the spark-plug gap or the contact-point gap.

gas A state of matter in which the matter has neither a definite shape nor a definite volume; air is a mixture of several gases. In an automobile, the discharge from the tail pipe is called the *exhaust gas*. Also, gas is a slang expression for the liquid fuel gasoline.

gasifier section That part of a gas-turbine engine which draws in the air, compresses it, mixes it with fuel, and burns the mixture in the combustor.

gasket A layer of material, usually made of cork or metal or both, that is placed between two machined surfaces to provide a tight seal between them.

gasket cement A liquid adhesive material, or sealer, used to install gaskets; in some applications, a layer of gasket cement is used as the gasket.

gasoline A liquid blend of hydrocarbons, obtained from crude oil; used as the fuel in most automobile engines.

gassing Hydrogen gas escaping from a battery; the gas is formed during battery charging.

gas-turbine engine A type of internal-combustion engine in which the shaft is spun by the pressure of combustion gases flowing against curved turbine blades located around the shaft.

gauge pressure A pressure read on a scale which ignores atmospheric pressure. Thus, the atmospheric pressure of 14.7 psi absolute is equivalent to 0 psi gauge.

gauge set One or more instruments attached to a manifold (a pipe fitted with several outlets for connecting pipes) and used for measuring pressure.

GCW Abbreviation for *gross combination weight;* the total weight of a tractor and semitrailer or trailer, including payload, fuel, driver, etc.

geared speed The theoretical vehicle speed; based on engine rpm, transmission-gear ratio, rear-axle ratio, and tire size.

gear lubricant A type of grease or oil designed especially to lubricate gears.

gear ratio The number of revolutions of a driving gear required to turn a driven gear through one complete revolution. For a pair of gears, the ratio is found by dividing the number of teeth on the driven gear by the number of teeth on the driving gear.

gear-type pump A pump in which a pair of rotating gears mesh to force oil (or some other liquid) from between the teeth to the pump outlet.

gears Mechanical devices that transmit power or turning effort from one shaft to another; gears contain teeth that mesh as the gears turn.

gearshift A linkage-type mechanism by which the gears in an automobile transmission are engaged.

generator A device that converts mechanical energy into electrical energy; can produce either ac or dc electricity. In automotive usage, a dc generator (now seldom used).

gingerbread Performance term for nonfunctional chrome and ornamentation.

glad hand The air-brake connector between a tractor and its trailer.

glaze The very smooth, mirrorlike finish that develops on engine-cylinder walls.

glaze breaker A tool, rotated by an electric motor, used to remove the glaze from engine-cylinder walls.

glow plug A plug-type heater containing a coil of resistance wire that is heated by a low-voltage current to ignite fuel sprayed into the intake manifold; used as a cold-starting aid.

goggles Special glasses worn over the eyes to protect them from flying chips, dirt, dust, and spraying refrigerant.

governor A device that controls, or governs, another device, usually on the basis of speed or rpm. For example, the governor used in certain automatic transmissions to control gear-shifting in relation to car speed.

GP Performance-term abbreviation for *Grand Prix*.

gradeability Ability of a truck to negotiate a given grade at a specified GCW or GVW.

grams per mile Unit of measurement for the amount (weight) of pollutants emitted into the atmosphere with the vehicle exhaust gases. Antipollution laws set maximum limits for each exhaust pollutant in grams per mile.

grease Lubricating oil to which thickening agents have been added.

greasy friction The friction between two solids coated with a thin film of oil.

grinder A machine for removing metal by means of a rotating abrasive wheel or stone.

grinding wheel A wheel made of abrasive material; used for removing metal from objects held against it while it rotates.

grommet A device, usually made of hard rubber or a like material, used to encircle or support a component. In emission systems, a grommet is located in the valve-cover assembly to support and help seal the PCV valve.

gross torque The maximum turning effort developed by an engine with optimal ignition setting, without allowances for the power absorbed by the engine's fan, alternator, water pump, and exhaust system.

grounding Connection of an electric unit to the vehicle engine or frame to provide a return path for electric current.

ground-return system Common system of electric wiring in which the chassis and frame of a vehicle are used as part of the electric return circuit to the battery or alternator; also known as the *single-wire system*.

growler An electrical test instrument for checking starting motors and generator armatures.

GT Performance-term abbreviation for *grand turismo*.

Guide-Matic An electronic device that automatically controls the headlights, shifting between upper and lower beams as conditions require.

guide sleeve A tubular sleeve that is put on a connecting-rod bolt when the rod is removed, to prevent scratching of the crankpin by the bolt threads.

gulp valve In the air-injection system, a type of antibackfire valve which allows a sudden intake of fresh air through the intake manifold during deceleration; prevents backfiring and popping in the exhaust system.

GVW Abbreviation for *gross vehicle weight;* the total weight of a vehicle, including the body, payload, fuel, driver, etc.

gymkhana A competitive meet using timed contests to test driving skills in parking, backing, and avoiding obstacles.

H

hacksaw A saw with a thin removable blade; used to cut metals.

hand jack A small jack that can be carried by hand; usually hydraulically operated, and used to lift a corner of a vehicle.

harmonic balancer See **vibration damper**.

hazard system Also called the *emergency signal system;* a driver-controlled system of flashing front and rear lights, used to warn approaching motorists when a car has broken down.

HC Abbreviation for **hydrocarbon**.

head See **cylinder head**.

header Performance term for a special exhaust manifold or exhaust tubes.

head-land ring A compression ring with an L-shaped cross-section; used as the top compression ring.

headlights Lights at the front of a vehicle; designed to illuminate the road ahead of the vehicle.

head pressure Pressure at the air-conditioner compressor outlet (the discharge pressure).

heat A form of energy; released by the burning of fuel. In an engine, heat energy is converted to mechanical energy.

heat-control valve In the engine, a thermostatically operated valve in the exhaust manifold; diverts heat to the intake manifold to warm it before the engine reaches normal operating temperature.

heat dam In a piston, a groove cut out to reduce the size of the path through which heat can travel; allows the piston skirt to run cooler.

heated-air system A system in which a thermostatically controlled air cleaner supplies hot air from a stove around the exhaust manifold to the carburetor during warm-up; improves cold-engine operation.

heater core A small radiator, mounted under the dash, through which hot coolant circulates. When heat is needed in the passenger compartment, a fan is turned on to circulate air through the hot core. Also, any liquid-to-air heat exchanger.

heat of compression Increase of temperature brought about by the compression of air or air-fuel mixture.

heat sink A device for absorbing heat from one medium and transferring it to another. The diodes in alternators are usually mounted in heat sinks which remove the heat from the diodes and thus prevent them from overheating.

HEI See High-Energy Ignition (HEI) System.

helical gear A gear in which the teeth are cut at an angle to the center line of the gear.

Heli-Coil A rethreading device used when threads are worn or damaged. The device is installed in a retapped hole to reduce the thread size to the original size.

hemi Performance term for any engine with a hemispherical combustion chamber.

hemispherical combustion chamber A combustion chamber resembling a hemisphere, or half a round ball.

hesitation Momentary pause in the rate of acceleration; momentary lack of throttle response at some car speed other than acceleration from a standing start.

high-compression Term used to refer to the increased compression ratios of modern automotive engines, as compared to engines built in the past.

high-discharge test A battery test in which the battery is discharged at a high rate while the cell voltages are checked.

high-energy ignition system A General Motors electronic ignition system without contact points, and with all ignition-system components contained in the distributor. Capable of producing 35,000 volts.

high-load condition Situation in which an air conditioner must operate continuously at maximum capacity to provide the required cooling; for example, at high temperature and high humidity.

high-pressure lines The lines from the air-conditioner compressor outlet to the thermostatic-expansion-valve inlet that carry high-pressure refrigerant. The two longest high-pressure lines are the discharge and liquid lines.

high-pressure vapor line Same as **discharge line.**

high side Same as **discharge side.**

high-side pressure Same as **discharge pressure.**

high-speed system In the carburetor, the system that supplies fuel to the engine at speeds above about 25 mph [40 km/h]. Also called the *main metering system.*

high-voltage cables The secondary (or spark-plug) cables or wires that carry high voltage from the ignition coil to the spark plugs.

hoist See **car lift.**

Honda system A type of controlled-combustion system for spark-ignition engines. Has a small chamber that surrounds the spark-plug electrodes with a rich mixture; once the rich mixture ignites, it enters the main chamber, igniting the leaner air-fuel mixture in that chamber.

hone An abrasive stone that is rotated in a bore or bushing to remove material.

hood The part of the car body that fits over and protects the engine.

horn An electrical noise-making device on a vehicle; used for signaling.

horn relay A relay connected between the battery and the horns. When the horn button is pressed, the relay is energized; it then connects the horns to the battery.

horsepower A measure of mechanical power, or the rate at which work is done. One horsepower equals 33,000 ft-lb (foot-pounds) of work per minute; it is the power necessary to raise 33,000 lb a distance of 1 ft in 1 minute.

H_2O Chemical symbol for hydrogen oxide.

Hotchkiss drive The type of rear suspension in which leaf springs absorb the rear-axle-housing torque.

hot-idle compensator A thermostatically controlled carburetor valve that opens whenever inlet air temperatures are high. Allows additional air to discharge below the throttle plates at engine idle to improve idle stability and prevent overly rich air-fuel mixtures.

hot patching A method of repairing a tire or tube by using heat to vulcanize a patch onto the damaged surface.

hot soak A condition that may arise when an engine is stopped for a prolonged period after a hard, hot run. Heat transferred from the engine evaporates fuel out of the carburetor, so that the carburetor needs priming before the engine will start and run smoothly. Requires a longer cranking period.

hot tank See **boil tank.**

hub The center part of a wheel.

humidity A measure of the amount of water vapor in the air.

hydraulic brakes A braking system that uses hydraulic pressure to force the brake shoes against the brake drums, or rotors, as the brake pedal is depressed.

hydraulic clutch A clutch that is actuated by hydraulic pressure; used in heavy-duty equipment, and where the engine is some distance from the driver's compartment so it would be difficult to use mechanical linkages.

hydraulic press A piece of shop equipment that develops a heavy force by use of a hydraulic piston-and-jack assembly.

hydraulic pressure Pressure exerted through the medium of a liquid.

hydraulics The use of a liquid under pressure to transfer force or motion, or to increase an applied force.

hydraulic valve A valve in a hydraulic system that operates on, or controls, the hydraulic pressure in the system. Also, any valve that is operated or controlled by hydraulic pressure.

hydraulic valve lifter A valve lifter that, by means of oil pressure, maintains zero valve clearance so that valve noise is reduced.

hydro Performance term for an automatic transmission.

hydrocarbon (HC) An organic compound containing only carbon and hydrogen, usually derived from fossil fuels such as petroleum, natural gas, and coal; an agent in the formation of photochemical smog. Gasoline is a blend of liquid hydrocarbons refined from crude oil.

hydrocarbon reactivity A measure of the smog-forming potential of a hydrocarbon.

hydrogen (H) A colorless, odorless, highly flammable gas whose combustion produces water; the simplest and lightest element.

hydrometer A device used to measure specific gravity. In automotive servicing, a device used to measure the specific gravity of battery electrolyte to determine the state of the battery charge; also a device used to measure the specific gravity of coolant to determine its freezing temperature.

hypoid gear A type of gear used in the differential (drive pinion and ring gear); cut in a spiral form to allow the pinion to be set below the center line of the ring gear, so that the car floor can be designed lower.

I

IC See **internal combustion engine**.

idle Engine speed when the accelerator pedal is fully released, and there is no load on the engine.

idle limiter A device that controls the maximum richness of the idle air-fuel mixture in the carburetor; also aids in preventing overly rich idle adjustments. Limiters are of two types: the external plastic-cap type, installed on the head of the idle-mixture adjustment screw, and the internal-needle type, located in the idle passages of the carburetor.

idle-limiter cap A plastic cap placed over the head of the idle-mixture adjustment screw, to limit its travel and prevent the idle mixture from being set too rich.

idle mixture The air-fuel mixture supplied to the engine during idling.

idle-mixture adjustment screw The adjustment screw (on some carburetors) that can be turned in or out to lean out or enrich the idle mixture.

idle port The opening into the throttle body through which the idle system in the carburetor discharges fuel.

idler arm In the steering system, a link that supports the tie rod and transmits steering motion to both wheels through the tie-rod ends.

idle speed The speed, or rpm, at which the engine runs without load when the accelerator pedal is released.

idle-stop solenoid An electrically operated two-position plunger used to provide a predetermined throttle setting at idle.

idle system In the carburetor, the passages through which fuel is fed when the engine is idling.

idle vent An opening from an enclosed chamber through which air can pass under idle conditions.

ignition The action of the spark in starting the burning of the compressed air-fuel mixture in the combustion chamber.

ignition advance The moving forward, in time, of the ignition spark relative to the piston position. TDC or 1° ATDC is considered advanced as compared to 2° ATDC.

ignition coil The ignition-system component that acts as a transformer to step up (increase) the battery voltage to many thousands of volts; the high-voltage surge from the coil is transmitted to the spark plug to ignite the compressed air-fuel mixture.

ignition distributor The ignition-system component that closes and opens the primary circuit to the ignition coil at the proper times and distributes the resulting high-voltage surges from the ignition coil to the proper spark plugs.

ignition reserve Difference between the minimum available and maximum required voltages. An adequate ignition reserve is important if an engine is to be reasonably free from troubles caused by moisture or dirt losses, leaky secondary leads, and fouled spark plugs.

ignition resistor A resistance connected into the ignition primary circuit to reduce the battery voltage to the coil during engine operation.

ignition retard The moving back, in time, of the ignition spark relative to the piston position. TDC or 1° BTDC is considered retarded as compared to 2° BTDC.

ignition switch The switch in the ignition system (usually operated with a key) that opens and closes the ignition-coil primary circuit. May also be used to open and close other vehicle electric circuits.

ignition system In the automobile, the system that furnishes high-voltage sparks to the engine cylinders to fire the compressed air-fuel mixture. Consists of the battery, ignition coil, ignition distributor, ignition switch, wiring, and spark plugs.

ignition timing The delivery of the spark from the coil to the spark plug at the proper time for the power stroke, relative to the piston position.

I-head engine An overhead-valve (OHV) engine; an engine with the valves in the cylinder head.

ihp Abbreviation for **indicated horsepower.**

IMCO Abbreviation for *improved combustion system,* an exhaust-emission control system used by Ford and comprised mainly of carburetor and distributor modifications. See also **controlled-combustion system.**

impact wrench An air-powered or electrically driven hand-held tool that rapidly turns nuts and bolts using a series of sharp fast blows, or impacts.

impeller A rotating finned disk; used in centrifugal pumps, such as water pumps, and in torque converters.

included angle In the front-suspension system, camber angle plus steering-axis inclination angle.

independent front suspension The conventional front-suspension system in which each front wheel is independently supported by a spring.

indicated horsepower The power produced within the engine cylinders, before any friction loss is deducted.

indicator A device used to make some condition known by use of a light or a dial and pointer; for example, the temperature indicator or oil-pressure indicator.

induction The action of producing a voltage in a conductor or coil by moving the conductor or coil through a magnetic field, or by moving the field past the conductor or coil.

inductive-type semiconductor ignition system An ignition system in which the primary energy is stored in an inductor or coil. This is the type now used in Chrysler, Ford, and General Motors HEI systems.

Indy Performance term for the Indianapolis 500-mile race.

inertia Property of an object that causes it to resist any change in its speed or the direction of its travel.

infrared analyzer A nondispersive test instrument used to measure very small quantities of pollutants in exhaust gas. See **exhaust-gas analyzer.**

injector The tube or nozzle through which fuel is injected into the intake airstream or the combustion chamber. Also, a performance term used for an engine equipped with fuel injection.

in-line engine An engine in which all the cylinders are located in a single row or line.

in-line steering gear A type of integral power steering; uses a recirculating-ball steering gear to which are added a control valve and an actuating piston.

inner tube See **tire tube.**

inside micrometer A precision tool used to measure the inside diameter of a hole.

inspect To examine a component or system for surface, condition, or function.

install To set up for use on a vehicle any part, accessory, option, or kit.

insulated-return system System of vehicle electrical wiring in which a separate insulated wire is used to provide the electric return circuit to the battery or alternator; also known as the *two-wire system.*

insulation Material that stops the travel of electricity (electrical insulation) or heat (heat insulation).

insulator A poor conductor of electricity or of heat.

intake manifold A device with several passages through which the air-fuel mixture flows from the carburetor to the ports in the cylinder head or cylinder block.

intake stroke The piston stroke from TDC to BDC immediately following the exhaust stroke, during which the intake valve opens and the cylinder fills with air-fuel mixture from the intake manifold.

intake valve The valve that opens during the intake stroke to allow the air-fuel mixture to enter the cylinder from the intake manifold.

integral Built into, as part of the whole.

inter-axle differential A two-position differential, located between the two driving axles of a tandem axle. In the unlocked position it divides the power unevenly between two axles and permits one axle to turn faster the other. In the locked position the power is divided approximately evenly between the two axles.

interchangeability The manufacture of similar parts to close tolerances so that any one of the parts can be substituted for another in a device, and the part will fit and operate properly; the basis of mass production.

internal-combustion (IC) engine An engine in which the fuel is burned inside the engine itself, rather than in a separate device (as is the case for a steam engine).

internal gear A gear with teeth pointing inward, toward the hollow center of the gear.

J

jack stand A pinned, or locked, type of safety stand placed under a car to support its weight after the car has been raised with a floor jack. Also called a *car stand* or *safety stand*.

jet A calibrated passage, in the carburetor, through which fuel flows.

journal The part of a rotating shaft which turns in a bearing.
jug Performance term for carburetor.

K

Kettering ignition system An inductive ignition system commonly used for automobile engines; employs an induction coil, breaker contacts (points), a capacitor (condenser), and a power supply such as a battery. Also called the *conventional* or *breaker-point* ignition system.

key A wedgelike metal piece, usually rectangular or semicircular, inserted in grooves to transmit torque while holding two parts in the same relative position. Also, the small strip of metal with coded peaks and grooves used to operate a lock, such as that for the ignition switch.

kickdown In automatic transmissions, a system that produces a downshift when the accelerator is pushed down to the floorboard.

kilogram (kg) In the metric system, a unit of weight and mass; approximately equal to 2.2 lb.

kilometer (km) In the metric system, a unit of linear measure; equal to 0.621 mi.

kilowatt (kW) A unit of power, equal to about 1.34 hp.

kinetic energy The energy of motion; the energy stored in a moving body through its momentum; for example, the kinetic energy stored in a rotating flywheel.

kingpin In older cars and trucks, the steel pin on which the steering knuckle pivots; attaches the steering knuckle to the knuckle support or axle.

kingpin inclination Inward tilt of the kingpin from the vertical. See **steering-axis inclination.**

knock A heavy metallic engine sound which varies with engine speed; usually caused by a loose or worn bearing. Name also used for detonation, pinging, and spark knock. See **detonation.**

knock-off Performance term for a special wheel lug designed to be removed quickly for fast tire changing.

knuckle A steering knuckle; a front-suspension part that acts as a hinge to support a front wheel and permit it to be turned to steer the car. The knuckle pivots on ball joints or, in earlier models, on kingpins.

knurl A series of ridges, formed on the outer surface of a piston or on the inner surface of a valve guide; helps reduce excess clearance and hold oil for added lubrication and quieter operation.

kW Abbreviation for **kilowatt.**

L

laminated Made up of several thin sheets or layers.

lamp A divisible assembly that provides light; contains a bulb or other light source and sometimes a lens and reflector.

landing gear Two small wheels at the forward end of a semitrailer; used to support the trailer when it is detached from the tractor.

lapping A method of seating engine valves in which the valve is turned back and forth on the seat; no longer recommended by car manufacturers.

lash The amount of free motion in a gear train, between gears, or in a mechanical assembly, such as the lash in a valve train.

latent heat "Hidden" heat required to change the state of a substance without changing its temperature. Latent heat cannot be felt or measured with a thermometer.

lead A cable or conductor to carry electric current (pronounced *leed*). A heavy metal; used in lead-acid storage batteries.

leaded gasoline Gasoline to which small amounts of tetraethyl lead are added to improve engine performance and reduce detonation.

lead sulfate A hard, insoluble layer that slowly forms on the plates of a discharging battery, and which may be reduced only by slow charging.

leaf spring A spring made up of a single flat steel plate, or several plates of graduated lengths assembled one on top of another; used on vehicles to absorb road shocks by bending, or flexing, in the middle.

leak detector Any device used to locate an opening where refrigerant may escape. Common types are flame, electronic, dye, and soap bubbles.

lean mixture An air-fuel mixture that has a relatively high proportion of air and a relatively low proportion of fuel. An air-fuel ratio of 16:1 indicates a lean mixture, compared to an air-fuel ratio of 13:1.

L-head engine An engine with its valves located in the cylinder block.

lift See **car lift.**

lifter See **valve lifter.**

lifter foot See **valve-lifter foot.**

light A gas-filled bulb enclosing a wire that glows brightly when an electric current passes through it; a lamp. Also, any visible radiant energy.

light-duty vehicle Any motor vehicle manufactured primarily for transporting persons or property and having a gross vehicle weight of 6,000 lb [2,727.6 kg] or less.

limited-slip differential A differential designed so that, when one wheel is slipping, a major portion of the drive torque is supplied to the wheel with the better traction; also called a *nonslip differential.*

linear measurement A measurement taken in a straight line; for example, the measurement of crankshaft end play.

line boring Using a special boring machine, centered on the original center of the cylinder-block main-bearing bores, to rebore the crankcase into alignment.

lines of force See **magnetic lines of force.**

lining See **brake lining.**

linkage An assembly of rods, or links, used to transmit motion.

linkage-type power steering A type of power steering in which the power-steering units (power cylinder and valve) are part of the steering linkage; frequently a bolt-on type of unit.

liquefied petroleum gas (LPG) A hydrocarbon suitable for use as an engine fuel, obtained from petroleum and natural gas; a vapor at atmospheric pressure but becomes a liquid under sufficient pressure. Butane and propane are the liquefied gases most frequently used in automotive engines.

liquid-cooled engine An engine that is cooled by the circulation of liquid coolant around the cylinders.

liquid line In an air conditioner, hose that connects the receiver-dehydrator outlet and the thermostatic-expansion-valve inlet. High-pressure liquid refrigerant flows through the line.

liter (l) In the metric system, a measure of volume; approximately equal to 0.26 gal (U.S.), or about 61 in^3. Used as a metric measure of engine-cylinder displacement.

live axle An axle that drives wheels which are rigidly attached to it.

loading An enrichment of the air-fuel mixture to the point of rough engine idle; sometimes causes missing and is usually accompanied by the emission of black smoke from the tail pipe.

load test A starting-motor test in which the current draw is measured under normal cranking load.

lobe A projecting part; for example, the rotor lobe or the cam lobe.

locknut A second nut turned down on a holding nut to prevent loosening.

lockwasher A type of washer which, when placed under the head of a bolt or nut, prevents the bolt or nut from working loose.

lower beam A headlight beam intended to illuminate the road ahead of the vehicle when meeting or following another vehicle.

low-lead fuel Gasoline which is low in tetraethyl lead, containing not more than 0.5 g per gallon.

low-pressure line Same as **suction line.**

low-pressure vapor line Same as **suction line.**

low side Same as **suction side.**

low-speed system The system in the carburetor that supplies fuel to the air passing through during low-speed, part-throttle operation.

LPG Abbreviation for **liquefied petroleum gas.**

lubricating system The system in the engine that supplies engine parts with lubricating oil to prevent actual contact between any two moving metal surfaces.

lugging Low-speed, full-throttle engine operation in which the engine is heavily loaded and overworked; usually caused by failure of the driver to shift to a lower gear when necessary.

M

machining The process of using a machine to remove metal from a metal part.

Magna-Flux A process in which an electromagnet and a special magnetic powder are used to detect cracks, in iron and steel, which might otherwise be invisible to the naked eye.

magnetic Having the ability to attract iron. This ability may be permanent, or it may depend on a current flow, as in an electromagnet.

magnetic clutch A magnetically-operated clutch used to engage and disengage the air-conditioner compressor.

magnetic field The area (or field) of influence of a magnet, within which it will exhibit magnetic properties; extends from the "north" pole of the magnet to its "south" pole. The strength of the field of an electromagnet increases with the number of turns of wire around the iron core and the current flow through the wire.

magnetic lines of force The imaginary lines by which a magnetic field may be visualized.

magnetic pole The point where magnetic lines of force enter or leave a magnet.

magnetic switch A switch with a winding (a coil of wire); when the winding is energized, the switch is moved to open or close a circuit.

magnetism The ability, either natural or produced by a flow of electric current, to attract iron.

magneto An engine-driven device that generates its own primary current, transforms that current into high-voltage surges, and delivers them to the proper spark plugs.

mag wheel A magnesium wheel assembly; also used to refer to any chromed, aluminum offset, or wide-rim wheel of spoke design.

main bearings In the engine, the bearings that support the crankshaft.

main jet The fuel nozzle, or jet, in the carburetor that supplies fuel when the throttle is partially to fully open.

make A distinctive name applied to a group of vehicles produced by one manufacturer; may be further subdivided into carlines, body types, etc.

malfunction Improper or incorrect operation.

manifold A device with several inlet or outlet passageways through which a gas or liquid is gathered or distributed. See **exhaust manifold, intake manifold, manifold gauge set.**

manifold gauge set A high-pressure and a low-pressure gauge mounted together as a set, used for checking pressures in the air-conditioning system.

manifold heat control See **heat-control valve.**

manifold vacuum The vacuum in the intake manifold that develops as a result of the vacuum in the cylinders on their intake strokes.

manual low Position of the units in an automatic transmission when the driver moves the shift lever to the low or first-gear position on the quadrant.

manufacturer Any person, firm, or corporation engaged in the production or assembly of motor vehicles or other products.

mass production The manufacture of interchangeable parts and similar products in large quantities.

master cylinder The liquid-filled cylinder, in the hydraulic braking system or clutch, where hydraulic pressure is developed when the driver depresses a foot pedal.

matter Anything that has weight and occupies space.

measuring The act of determining the size, capacity, or quantity of an object.

mechanical efficiency In an engine, the ratio between brake horsepower and indicated horsepower.

mechanism A system of interrelated parts that make up a working assembly.

member Any essential part of a machine or assembly.

meshing The mating, or engaging, of the teeth of two gears.

meter (m) A unit of linear measure in the metric system, equal to 39.37 in. Also, the name given to any test instrument that measures a property of a substance passing through it, as an ammeter measures electric current. Also, any device that measures and controls the flow of a substance passing through it, as a carburetor jet meters fuel flow.

metering oil pump In a Mazda Wankel rotary engine, a plunger-type pump, driven by a distributor drive gear, which meters oil to the carburetor for mixing with the gasoline. In the combustion chamber, this small amount of oil lubricates the rotor seals.

metering rod and jet A device consisting of a small, movable, cone-shaped rod and a jet; increases or decreases fuel flow according to engine throttle opening, engine load, or a combination of both.

metering valve A valve in the disk-brake system which prevents hydraulic pressure to the front brakes until after the rear brakes are applied, and thus prevents rear-end skidding.

mica An insulating material used to separate the copper bars of commutators.

micrometer A precision measuring device used to measure small bores, diameters, and thicknesses. Also called a "mike."

mike Slang term for **micrometer.**

millimeter (mm) In the metric system, a unit of linear measure, approximately equal to 0.039 in.

misfire In the engine, a failure to ignite the air-fuel mixture in one or more cylinders. This condition may be intermittent or continuous in one or more cylinders.

miss See **misfire.**

mode Term used to designate a particular set of operating characteristics.

model year The production period for new motor vehicles or new engines, designated by the calendar year in which the period ends.

modification An alteration; a change from the original.

modulator A pressure-regulated governing device; used, for example, in automatic transmissions.

moisture Humidity, dampness, wetness, or very small drops of water.

mold A hollow form into which molten metal is poured and allowed to harden.

molecule The smallest particle into which a substance can be divided and still retain the properties of the substance.

MON Abbreviation for **motor octane number.**

monolithic Made as a single unit. In catalytic-converter construction, a substrate or supporting structure for the catalyst, made as a single unit (usually in the shape of a honeycomb), is monolithic; however, the coated-bead or pellet-type catalytic converter is not.

monolithic timing Making accurate spark-timing adjustments with an electronic timing device which can be used with the engine running.

motor A device that converts electric energy into mechanical energy; for example, the starting motor.

Motor Octane Number (MON) Laboratory octane rating of a fuel, established on single-cylinder, variable-compression-ratio engines.

motor vehicle A vehicle propelled by a means other than muscle power, usually mounted on rubber tires, which does not run on rails or tracks.

mph Abbreviation for *miles per hour,* a unit of speed.

muffler In the engine exhaust system, a device through which the exhaust gases must pass and which reduces the exhaust noise. In an air-conditioning system, a device to minimize pumping sounds from the compressor.

multiple-disk clutch A clutch with more than one friction disk; usually, there are several driving disks and several driven disks, alternately placed.

multiple-viscosity oil An engine oil which has a low viscosity when cold (for easier cranking) and a higher viscosity when hot (to provide adequate engine lubrication).

muscle car Performance term for a car with a low weight-to-horsepower ratio.

mushroomed valve stem A valve stem that has worn so that the tip, or butt end, has spread (mushroomed) and metal is hanging over the valve guide. The valve cannot be removed from the valve guide until the mushroomed metal is removed.

mutual induction The condition in which a voltage is induced in one coil by a changing magnetic field caused by a changing current in another coil. The magnitude of the induced voltage depends on the number of turns in the two coils.

N

NASCAR Abbreviation for *National Association for Stock Car Auto Racing.*

neck A portion of a shaft that has a smaller diameter than the rest of the shaft.

needle bearing An antifriction bearing of the roller type, in which the rollers are very small in diameter (needle-sized).

needle valve A small, tapered, needle-pointed valve which can move into or out of a valve seat to close or open the passage through the seat. Used to control the carburetor float-bowl fuel level.

negative One of the two poles of a magnet, or one of the two terminals of an electrical device.

negative terminal The terminal from which electrons flow in a complete electric circuit. On a battery, the negative terminal can be identified as the battery post with the smaller diameter; the minus sign ($-$) is often also used to identify the negative terminal.

neoprene A synthetic rubber that is not affected by the various chemicals that are harmful to natural rubber.

net torque The turning effort developed by an engine with optimum ignition setting, after allowances for the power absorbed by the engine's fan, alternator, water pump, and exhaust system.

neutral In a transmission, the setting in which all gears are disengaged, and the output shaft is disconnected from the drive wheels.

neutral-start switch A switch wired into the ignition switch to prevent engine cranking unless the transmission shift lever is in NEUTRAL.

neutron A particle in the nucleus of an atom; weighs about the same as a proton, but has no electric charge.

NHTSA Abbreviation for *National Highway Traffic Safety Administration.*

nitro Performance term for nitromethane, a fuel additive used to obtain greater power from a racing engine.

nitrogen (N) A colorless, tasteless, odorless gas that constitutes 78 percent of the atmosphere by volume and is a part of all living tissues.

nitrogen oxides (NO_x) Any chemical compound of nitrogen and oxygen. Nitrogen oxides result from high temperature and pressure in the combustion chambers of automobile engines and other power plants during the combustion process. When combined with hydrocarbons in the presence of sunlight, nitrogen oxides form smog. A basic air pollutant; automotive exhaust-emission levels of nitrogen oxides are controlled by law.

noble metals Metals (such as gold, silver, platinum, and palladium) which do not readily oxidize or enter into other chemical reactions, but do promote reactions between other substances. Platinum and palladium are used as catalysts in catalytic converters.

no-load test A starting-motor test in which the starting motor is operated without load, and the current draw and armature speed at specified voltages are noted.

nonconductor Same as **insulator.**

nonleaded gasoline See **unleaded gasoline.**

nonslip differential See **limited-slip differential.**

north pole The pole from which the lines of force leave a magnet.

NO_x Abbreviation for any **nitrogen oxides.**

NO_x control system Any device or system used to reduce the amount of NO_x produced by an engine.

nozzle The opening, or jet, through which fuel passes when it is discharged into the carburetor venturi.

nucleus The central part of an atom; it has a positive charge (due to the protons).

nut A removable fastener used with a bolt to lock pieces together; made by threading a hole through the center of a piece of metal which has been shaped to a standard size.

O

octane number The number used to indicate the octane rating of a gasoline.

octane rating A measure of the antiknock properties of a gasoline. The higher the octane rating, the more resistant the gasoline is to spark knock or detonation.

octane requirement The minimum-octane-number fuel required to enable a vehicle to operate without knocking.

odometer The meter that indicates the total distance a vehicle has traveled, in miles or kilometers; usually located in the speedometer.

OEM Abbreviation for **original-equipment manufacturer.**

OHC See **overhead-camshaft engine.**

ohm The unit of electrical resistance.

ohmmeter An instrument used to measure electrical resistance.

OHV See **overhead-valve engine.**

oil A liquid lubricant; made from crude oil and used to provide lubrication between moving parts. In a diesel engine, oil is used for fuel.

oil bleed line In an air conditioner, a passageway that ensures positive oil return to the compressor at high compressor speeds and under low charge conditions.

oil-control ring See **oil ring.**

oil cooler A small radiator which lowers the temperature of oil flowing through it.

oil dilution Thinning of oil in the crankcase; caused by liquid gasoline leaking past the piston rings from the combustion chamber.

oil filter A filter which removes impurities from crankcase oil passing through it.

oil-injection cylinder A cylinder with which a measured quantity of refrigerant oil is added to an air-conditioning system or component.

oil-level indicator The indicator that is removed and inspected to check the level of oil in the crankcase of an engine or compressor. Usually called the *dipstick*.

oil pan The detachable lower part of the engine, made of sheet metal, which encloses the crankcase and acts as an oil reservoir.

oil-pressure indicator A gauge that indicates (to the driver) the oil pressure in the engine lubricating system.

oil pump In the lubricating system, the device that delivers oil from the oil pan to the moving engine parts.

oil pumping Leakage of oil past the piston rings and into the combustion chamber; usually the result of defective rings or worn cylinder walls.

oil ring The lower ring or rings on a piston; designed to prevent excessive amounts of oil from working up the cylinder walls and into the combustion chamber. Also called an *oil-control ring*.

oil seal A seal placed around a rotating shaft or other moving part to prevent leakage of oil.

oil seal and shield Two devices used to control oil leakage past the valve stem and guide, and into the ports or the combustion chamber of an engine.

oil separator A device for separating oil from air or from another liquid. Used with some engine-crankcase emission-control systems.

oil strainer A wire-mesh screen placed at the inlet end of the oil-pump pickup tube; prevents dirt and other large particles from entering the oil pump.

one-way clutch See **sprag.**

one-wire system On automobiles, use of the car body, engine, and frame as a path for the grounded side of the electric circuits; eliminates the need for a second wire as a return path to the battery or alternator.

open circuit In an electric circuit, a break, or opening, which prevents the passage of current.

open system A crankcase emission-control system which

draws air through the oil-filler cap, and does not include a tube from the crankcase to the air cleaner.

operational test Same as **performance test.**

orifice A small opening, or hole, into a cavity.

orifice spark-advance control A system used on some engines to aid in the control of NO_x. Consists of a valve which delays the change in vacuum to the distributor vacuum-advance unit between idle and part throttle.

O ring A type of sealing ring, made of a special rubberlike material; in use, the O ring is compressed into a groove to provide the sealing action.

OSAC Abbreviation for **orifice spark-advance control.**

oscillating Moving back and forth, as a swinging pendulum.

oscilloscope A high-speed voltmeter which visually displays voltage variations on a television-type picture tube. Widely used to check engine ignition systems; can also be used to check charging systems and electronic fuel-injection systems.

Otto cycle The cycle of events in a four-stroke-cycle engine. Named for the inventor, Dr. Nikolaus Otto.

output shaft The main shaft of the transmission; the shaft that delivers torque from the transmission to the drive shaft.

overcenter spring A spring used in some clutch linkages to reduce the foot pressure required to depress the clutch pedal.

overcharging Continued charging of a battery after it has reached the charged condition. This action damages the battery and shortens its life.

overdrive A device, usually mounted behind the transmission, which introduces an extra set of gears into the power train. This causes the drive shaft to overdrive, or drive faster than, the engine crankshaft.

overflow Spilling of the excess of a substance; also, to run or spill over the sides of a container, usually because of overfilling.

overflow tank See **expansion tank.**

overhaul To completely disassemble a unit, clean and inspect all parts, reassemble it with the original or new parts, and make all adjustments necessary for proper operation.

overhead-camshaft (OHC) engine An engine in which the camshaft is mounted over the cylinder head, instead of inside the cylinder block.

overhead-valve (OHV) engine An engine in which the valves are mounted in the cylinder head above the combustion chamber, instead of in the cylinder block; in this type of engine, the camshaft is usually mounted in the cylinder block, and the valves are actuated by pushrods.

overheat To heat excessively; also, to become excessively hot.

overrunning clutch drive A type of clutch drive which transmits rotary motion in one direction only; when rotary motion attempts to pass through in the other direction, the then driving member overruns and does not pass the motion to the other member. Widely used as the drive mechanism for starting motors.

oversquare Term applied to an automotive engine in which the bore is larger than the stroke.

oversteer A built-in characteristic of certain types of rear-suspension systems; causes the rear wheels to turn toward the outside of a turn.

oxidation Burning or combustion; the combining of a material with oxygen; rusting is slow oxidation, and combustion is rapid oxidation.

oxidation catalyst In a catalytic converter, a substance that promotes the combustion of exhaust-gas hydrocarbons and carbon monoxide at a lower temperature.

oxides of nitrogen See **nitrogen oxides.**

oxygen (O) A colorless, tasteless, odorless, gaseous element which makes up about 21 percent of air. Capable of combining rapidly with all elements except the inert gases in the oxidation process called *burning*. Combines very slowly with many metals in the oxidation process called *rusting*.

P

pan See **oil pan.**
pancake engine An engine with two rows of cylinders which are opposed and on the same plane; usually set in a

car horizontally. Examples are the Chevrolet Corvair and Volkswagen engines.

parade pattern An oscilloscope pattern showing the ignition voltages on one line, from left to right across the scope screen in engine firing order.

parallel The quality of being the same distance from each other at all points; usually applied to lines and, in automotive work, to machined surfaces.

parallel circuit The electric circuit formed when two or more electrical devices have their terminals connected together, positive to positive and negative to negative, so that each may operate independently of the others, from the same power source.

parallelogram linkage A steering system in which a short idler arm is mounted on the right side, so that it is parallel to the pitman arm.

parking brake Mechanically operated brake that is independent of the foot-operated service brakes on the vehicle; set when the vehicle is parked.

particle A very small piece of metal, dirt, or other impurity which may be contained in the air, fuel, or lubricating oil used in an engine.

particulates Small particles of lead occurring as solid matter in the exhaust gas.

passage A small hole or gallery in an assembly or casting, through which air, coolant, fuel, or oil flows.

passenger car Any four-wheeled motor vehicle manufactured primarily for use on streets and highways and carrying 10 passengers or fewer.

pawl An arm, pivoted so that its free end can fit into a detent, slot, or groove at certain times to hold a part stationary.

payload The weight of the cargo carried by a truck, not including the weight of the body.

PCV Abbreviation for **positive crankcase ventilation.**

PCV valve The valve that controls the flow of crankcase vapors in accordance with ventilation requirements for different engine speeds and loads.

pedal reserve The distance from the brake pedal to the floorboard after the brakes are applied.

peen To mushroom or spread the end of a pin or rivet.

percent of grade The quotient obtained by dividing the height of a hill by its length; used in computing the power requirements of trucks.

percolation The condition in which a bowl vent fails to open when the engine is turned off, and pressure in the fuel bowl forces raw fuel through the main jets into the manifold.

performance test The use of a manifold gauge set to measure the pressures in an air-conditioning system as a check of system performance.

permanent magnet A piece of steel that retains its magnetism without the use of an electric current to create a magnetic field.

petroleum The crude oil from which gasoline, lubricating oil, and other such products are refined.

photochemical smog Smog caused by hydrocarbons and nitrogen oxides reacting photochemically in the atmosphere. The reactions take place under low wind velocity, bright sunlight, and an inversion layer in which the air mass is trapped (as between the ocean and mountains in Los Angeles). Can cause eye and lung irritation.

pickup coil In an electronic ignition system, the coil in which voltage is induced by the reluctor.

pilot bearing A small bearing, in the center of the flywheel end of the crankshaft, which carries the forward end of the clutch shaft.

pilot shaft A shaft that is used to align parts, and which is removed before final installation of the parts; a dummy shaft.

ping Engine "knock" that occurs only during acceleration. Usually associated with medium to heavy throttle acceleration or lugging at relatively low speeds, especially with a manual transmission. However, it may occur in higher speed ranges under heavy-load conditions. Caused by too much advance of ignition timing or low-octane fuel.

pinion gear The smaller of two meshing gears.

pintle hook A hook mounted on a truck or trailer and used to couple a full trailer to a truck.

piston A movable part, fitted to a cylinder, which can receive or transmit motion as a result of pressure changes in a fluid.

In the engine, the cylindrical part that moves up and down within a cylinder as the crankshaft rotates.

piston crown The top of the piston.

piston displacement The cylinder volume displaced by the piston as it moves from the bottom to the top of the cylinder during one complete stroke.

piston pin The cylindrical or tubular metal piece that attaches the piston to the connecting rod. Also called the *wrist pin*.

piston-ring compressor A tool used in engine-overhaul work to compress the piston rings inside the piston grooves so the piston-and-ring assembly may be installed in the engine cylinder.

piston rings Rings fitted into grooves in the piston. There are two types: compression rings for sealing the compression in the combustion chamber, and oil rings to scrape excessive oil off the cylinder wall. See **compression ring** and **oil ring.**

piston skirt The lower part of the piston, below the piston-pin hole.

piston slap A hollow, muffled, bell-like sound made by an excessively loose piston slapping the cylinder wall.

pitch The number of threads per inch on any threaded part.

pitman arm In the steering system, the arm that is connected between the steering-gear sector shaft and the steering linkage, or tie rod; it swings back and forth for steering as the steering wheel is turned.

pivot A pin or shaft upon which another part rests or turns.

planetary-gear system A gear set consisting of a central sun gear surrounded by two or more planet pinions which are, in turn, meshed with a ring (or internal) gear; used in overdrives and automatic transmissions.

planet carrier In a planetary-gear system, the carrier or bracket that contains the shaft upon which the planet pinion turns.

planet pinions In a planetary-gear system, the gears that mesh with, and revolve about, the sun gear; they also mesh with the ring (or internal) gear.

plastic gasket compound A plastic paste which can be squeezed out of a tube to make a gasket in any shape.

Plastigage A plastic material available in strips of various sizes; used to measure crankshaft main-bearing and connecting-rod-bearing clearances.

plate In a battery, a rectangular sheet of spongy lead. Sulfuric acid in the electrolyte chemically reacts with the lead to produce an electric current.

plate group In a battery, all the positive plates or all the negative plates for one cell, connected together electrically.

plenum A chamber containing air under pressure.

plenum blower assembly In an air-conditioning system, the assembly through which air passes on its way to the evaporator or heater core. It is located on the engine side of the fire wall and contains a blower, air ducts, air valves, and doors that permits selection of air from outside or inside the car.

pliers Hand-held pincerlike tools used for cutting and gripping, and available in many styles and sizes.

plies The layers of cord in a tire casing, each of which is a ply.

pneumatic tool A tool powered by air, such as an air-powered impact wrench.

POA valve See **suction throttling valve.**

polarity The quality of an electrical component or circuit that determines the direction of current flow.

polarizing a generator Correcting the generator field polarity so the generator will build up polarity in the proper direction to charge the battery.

pole See **magnetic pole.**

pole shoe The curved metal shoe around which a field coil is placed.

pollutant Any substance that adds to the pollution of the atmosphere. In a vehicle, any such substance in the exhaust gas from the engine, or evaporating from the fuel tank or carburetor.

pollution Any gas or substance, in the air, which makes it less fit to breathe. Also, noise pollution is the name applied to excessive noise from machinery or vehicles.

polyurethane A synthetic substance used in filtration materials; normally associated with the filtering of carburetor inlet air.

pop-back   Condition in which the air-fuel mixture is ignited in the intake manifold. Because combustion takes place outside the combustion chamber, the combustion may "pop back" through the carburetor.

poppet valve   A mushroom-shaped valve, widely used in automotive engines.

port   In the engine, the opening in which the valve operates and through which air-fuel mixture or burned gases pass; the valve port.

ported vacuum switch   A water-temperature-sensing vacuum control valve used in distributor and EGR vacuum circuits. Sometimes called the *vacuum control valve* or *coolant override valve*.

positive   One of the two poles of a magnet, or one of the two terminals of an electrical device.

positive crankcase ventilation (PCV)   A crankcase ventilation system; uses intake-manifold vacuum to return the crankcase vapors and blow-by gases from the crankcase to the intake manifold to be burned, thereby preventing their escape into the atmosphere.

positive terminal   The terminal to which electrons flow in a complete electric circuit. On a battery, the positive terminal can be identified as the battery post with the larger diameter; the plus sign (+) is often also used to identify the positive terminal.

post   A point at which a cable is connected to the battery.

potential energy   Energy stored in a body because of its position. A weight raised to a height has potential energy because it can do work coming down. Likewise, a tensed or compressed spring contains potential energy.

pounds per horsepower   A measure of vehicle performance; the weight of the vehicle divided by the engine horsepower.

pour point   The lowest temperature at which an oil will flow.

power   The rate at which work is done. A common power unit is the horsepower, which is equal to 33,000 ft-lb/min (foot-pounds per minute).

power brakes A brake system that uses vacuum and atmospheric pressure to provide most of the effort required for braking.

power cylinder An operating cylinder which produces the power to actuate a mechanism. Both power brakes and power-steering units contain power cylinders.

power piston In some carburetors, a vacuum-operated piston that allows additional fuel to flow at wide-open throttle; permits delivery of a richer air-fuel mixture to the engine.

power plant The engine or power source of a vehicle.

power section In a gas-turbine engine, the section that contains the power-turbine rotors which, through reduction gears, turn the wheels of the vehicle.

power steering A steering system that uses hydraulic pressure (from a pump) to multiply the driver's steering effort.

power stroke The piston stroke from TDC to BDC immediately following the compression stroke, during which both valves are closed and the air-fuel mixture burns, expands, and forces the piston down to transmit power to the crankshaft.

power take-off An attachment for connecting the engine to devices or other machinery when its use is required.

power team The combination of an engine, transmission, and specific axle ratio.

power tool A tool whose power source is not muscle power; a tool powered by air or electricity.

power train The mechanisms that carry the rotary motion developed in the engine to the car wheels; includes the clutch, transmission, drive shaft, differential, and axles.

ppm Abbreviation for *parts per million;* the unit used in measuring the level of hydrocarbons in exhaust gas with an exhaust-gas analyzer.

PR Abbreviation for *ply rating;* a measure of the strength of a tire, based on the strength of a single ply of designated construction.

precision-insert bearings Bearings that can be installed in an engine without reaming, honing, or grinding.

precombustion chamber In some diesel engines, a separate small combustion chamber into which the fuel is injected and where combustion begins.

preignition Ignition of the air-fuel mixture in the combustion chamber, by any means, before the ignition spark occurs at the spark plug.

preload In bearings, the amount of load placed on a bearing before actual operating loads are imposed. Proper preloading requires bearing adjustment and assures alignment and minimum looseness in the system.

premium gasoline The best or highest-octane gas available to the motorist.

press fit A fit (between two parts) so tight that one part has to be pressed into the other, usually with an arbor press or hydraulic press.

pressure Force per unit area, or force divided by area. Usually measured in pounds per square inch (psi) and kilograms per square centimeter (kg/cm^2).

pressure bleeder A piece of shop equipment that uses air pressure to force brake fluid into the brake system for bleeding.

pressure cap A radiator cap, with valves, which causes the cooling system to operate under pressure and thus at a somewhat higher and more efficient temperature.

pressure-differential valve The valve in a dual-brake system that turns on a warning light if the pressure drops in one part of the system.

pressure-feed oil system A type of lubricating system that makes use of an oil pump to force oil to various engine parts.

pressure line See **discharge line.** In an air conditioner, all refrigerant lines are under pressure.

pressure plate That part of the clutch which exerts pressure against the friction disk; it is mounted on and rotates with the flywheel.

pressure regulator A device which operates to prevent excessive pressure from developing. In the hydraulic systems of certain automatic transmissions, a valve that opens to release oil from a line when the oil pressure reaches a specified maximum limit.

pressure-relief valve A valve in the oil line that opens to relieve excessive pressures.

pressure-sensing line In an air-conditioner, prevents the compressor suction pressure from dropping below a predetermined pressure; opens the thermostatic expansion valve, allowing liquid refrigerant to flood the evaporator.

pressure tester An instrument that clamps in the radiator filler neck; used to pressure-test the cooling system for leaks.

pressurize To apply more than atmospheric pressure to a gas or liquid.

preventive maintenance The systematic inspection of a vehicle to detect and correct failures, either before they occur or before they develop into major defects. A procedure for economically maintaining a vehicle in a satisfactory and dependable operating condition.

primary The low-voltage circuit of the ignition system.

primary winding The outer winding, of relatively heavy wire, in an ignition coil.

printed circuit An electric circuit made by applying a conductive material to an insulating board in a pattern that provides current paths between components mounted on or connected to the board.

PROCO Short for *programmed combustion;* a research type of stratified charge engine developed by Ford.

programmed combustion See **PROCO.**

programmed protection system A system employing bypass valves to protect the catalysts and their containers from destructive overtemperature conditions that might result from certain modes of operation, or from engine malfunctions.

progressive linkage A carburetor linkage used with multiple-carburetor installations to progressively open the secondary carburetors.

propane A type of LPG that is liquid below $-44°F$ [$-42°C$] at atmospheric pressure.

propeller shaft See **drive shaft.**

proportioning valve A valve which admits more braking pressure to the front wheels when high fluid pressures develop during braking.

proton A particle in the nucleus of an atom; has a positive electric charge.

prussian blue A blue pigment; in solution, useful in determining the area of contact between two surfaces.

psi Abbreviation for *pounds per square inch,* a unit of pressure.

psig Abbreviation for *pounds per square inch* of **gauge pressure.**

pull The result of an unbalanced condition. For example, uneven braking at the front brakes or unequal front-wheel alignment will cause a car to swerve (pull) to one side when the brakes are applied.

puller Generally, a shop tool used to separate two closely fitted parts without damage. Often contains a screw, or several screws, which can be turned to apply gradual pressure.

pulley A metal wheel with a V-shaped groove around the rim; drives, or is driven by, a belt.

pulsation A surge, felt in the brake pedal during low-pressure braking.

pump A device that transfers gas or liquid from one place to another.

punch A hand-held tool that is struck with a hammer to drive one piece of metal from inside another.

puncture-sealing tires or tubes Tires or tubes coated on the inside with a plastic material. Air pressure in the tire or tube forces the material into a puncture; it hardens on contact with the air to seal the puncture.

purge To remove, evacuate, or empty trapped substances from a space. In an air conditioner, to remove moisture and air from the refrigerant system by flushing with nitrogen or Refrigerant-12.

purge valve A valve used on some charcoal canisters, in evaporative emission control systems; limits the flow of vapor and air to the carburetor during idling.

pushrod In the I-head engine, the rod between the valve lifter and the rocker arm; transmits cam-lobe lift.

PVS Abbreviation for **ported vacuum switch.**

Q

quad Performance term for a four-barrel carburetor.

quad carburetor A four-barrel carburetor.

quadrant A term sometimes used to identify the shift-lever selector mounted on the steering column.

quarter Performance term for a quarter-mile drag strip.

quench The removal of heat during combustion from the end gas or outside layers of air-fuel mixture by the cooler metallic surfaces of the combustion chamber, thus reducing the tendency for detonation to occur.

quench area The area of the combustion chamber near the cylinder walls which tends to cool (quench) combustion through the effect of the nearby cool water jackets.

quick charger A battery charger which produces a high charging current and thus substantially charges, or boosts, a battery in a short time.

R

races The metal rings on which ball or roller bearings rotate.

rack-and-pinion steering gear A steering gear in which a pinion on the end of the steering shaft meshes with a rack on the major cross member of the steering linkage.

radial-ply tire A tire in which the plies are placed radially, or perpendicular to the rim, with a circumferential belt on top of them. The rubber tread is vulcanized on top of the belt and plies.

radiation One of the processes by which energy is transferred. For example, heat energy from the sun reaches the earth by radiation.

radiator In the cooling system, the device that removes heat from coolant passing through it; takes hot coolant from the engine and returns the coolant to the engine at a lower temperature.

radiator pressure cap See **pressure cap.**

radiator shutter system An engine-temperature control system, used mostly on trucks, that controls the amount of air flowing through the radiator by use of a shutter system.

radius ride Condition in which a crankshaft journal "rides" on the edge of the bearing. Caused by not cutting away enough of the radius of the journal, where it comes up to the crank cheek, when the crankshaft is reground.

rail job Performance term for a type of drag-racing car built on steel rails or tubes with very little body covering.

ram-air cleaner An air cleaner for high-performance cars that opens an air scoop on the hood, to provide a ram effect, when the throttle is wide open.

raster pattern An oscilloscope pattern showing the ignition voltages one above the other, from the bottom to the top of the screen in the engine firing order.

ratio Proportion; the relative amounts of two or more substances in a mixture. Usually expressed as a numerical relationship, as in 2:1.

RC engine Abbreviation for **rotary-combustion engine.** See **Wankel engine.**

reactor The stator in a torque converter; has reactive blades against which the fluid can change direction (under certain conditions) as it passes from the turbine to the pump.

reamer A round metal-cutting tool with a series of sharp cutting edges; enlarges a hole when turned in the hole.

rear-end torque The reaction torque that acts on the rear-axle housing when torque is applied to the wheels; tends to turn the axle housing in a direction opposite to wheel rotation.

reassembly Putting the parts of a device back together.

rebore To increase the diameter of a cylinder.

recapping A form of tire repair in which a cap of new tread material is placed on the old casing and vulcanized into place.

receiver In a car air conditioner, a metal tank for holding excess refrigerant. Liquid refrigerant is delivered from the condenser to the receiver.

receiver-dehydrator In a car air conditioner, a container for storing liquid refrigerant from the condenser. A sack of desiccant in this container removes small traces of moisture that may be left in the system after purging and evacuating.

recharging The action of forcing electric current into a battery in the direction opposite that in which current normally flows during use. Reverses the chemical reaction between the plates and electrolyte.

reciprocating motion Motion of an object between two limiting positions; motion in a straight line back and forth, or up and down.

recirculating-ball-and-nut steering gear A type of steering gear in which a nut (meshing with a gear sector) is assembled on a worm gear; balls circulate between the nut and worm threads.

rectifier A device which changes alternating current to direct current; in the alternator, a diode.

reed valve A type of valve used in the crankcases of some two-cycle engines. Air-fuel mixture enters the crankcase through the reed valve, which then closes as pressure builds up in the crankcase.

refractometer An instrument used to measure the specific gravity of a liquid such as battery electrolyte or engine coolant; gives a reading that is already adjusted for the temperature of the liquid being tested.

refrigerant A substance used to transfer heat in an air conditioner, through a cycle of evaporation and condensation.

Refrigerant-12 The refrigerant used in vehicle air-conditioning systems. It is sold under such trade names as Freon-12.

refrigeration Cooling an object or substance by removal of heat through mechanical means.

refrigeration cycle The complete sequence of changes (in temperature, pressure, and state) undergone by the refrigerant as it circulates through a refrigeration system.

regeneration system A system in a gas turbine that converts some of the heat that would otherwise be wasted into usable power.

regulator In the charging system, a device that controls alternator output to prevent excessive voltage.

relative humidity The actual moisture content of the air, as a percentage of the total moisture that the air can hold at a given temperature. For example, if the air contains three-fourths of the moisture it can hold at its existing temperature, then its relative humidity is 75 percent.

relay An electrical device that opens or closes a circuit or circuits in response to a voltage signal.

release bearing See **throwout bearing.**

release fingers See **release levers.**

release levers In the clutch, levers that are moved by throwout-bearing movement, causing clutch-spring pressure to be relieved so the clutch is released, or uncoupled from the flywheel.

relief valve A valve that opens when a preset pressure is reached. This relieves or prevents excessive pressures.

reluctor In an electronic ignition system, the metal rotor (with a series of tips) which replaces the conventional distributor cam.

remove and reinstall (R and R) To perform a series of servicing procedures on an original part or assembly; includes removal, inspection, lubrication, all necessary adjustments, and reinstallation.

replace To remove a used part or assembly and install a new part or assembly in its place; includes cleaning, lubricating, and adjusting as required.

required voltage The voltage required to fire a spark plug.

research octane number A number used to describe the octane rating of a marketed gasoline. See also **motor octane number.**

reserve capacity A battery rating; the number of minutes a battery can deliver a 25-A current before the cell voltages drop to 1.75 volts per cell.

residual magnetism The magnetism that remains in a material after the electric current producing the magnetism has stopped flowing.

resistance The opposition to a flow of current through a circuit or electrical device; measured in ohms. A voltage of 1 volt will cause 1 A to flow through a resistance of 1 ohm. This is known as Ohm's law, which can be written in three ways: amperes = volts/ohms; ohms = volts/amperes; and volts = amperes × ohms.

resonator A device in the exhaust system that reduces the exhaust noise.

retainer lock See **valve-spring-retainer lock.**

retard Usually associated with the spark-timing mechanisms of the engine: the opposite of spark advance. Also, to delay the introduction of the spark into the combustion chamber.

return spring A "pull-back" spring, often used in brake systems.

rev Performance term for revolutions per minute (rpm), or the speed of rotation of the engine crankshaft.

reverse flushing A method of cleaning a radiator or engine cooling system by flushing it in the direction opposite to normal coolant flow.

reverse idler gear In a transmission, an additional gear that must be meshed to obtain reverse gear; a gear used only in reverse, that idles when the transmission is in any other position.

ribbon-cellular radiator core A type of radiator core consisting of ribbons of metal soldered together along their edges.

rich mixture An air-fuel mixture that has a relatively high proportion of fuel and a relatively low proportion of air. An air-fuel ratio of 13:1 indicates a rich mixture, compared to an air-fuel ratio of 16:1.

ring See **compression ring** and **oil ring.**

ring expander A special tool used to expand piston rings for installation on the piston.

ring gap The gap between the ends of the piston ring when the ring is in place in the cylinder.

ring gear A large gear carried by the differential case; meshes with and is driven by the drive pinion.

ring grooves Grooves cut in a piston, into which the piston rings are assembled.

ring ridge The ridge left at the top of a cylinder as the cylinder wall below is worn away by piston-ring movement.

ring-ridge remover A special tool used to remove the ring ridge from a cylinder.

rivet A semipermanent fastener used to hold two pieces together.

roadability The steering and handling qualities of a vehicle while it is being driven on the road.

road-draft tube A method of scavenging the engine crankcase of fumes and pressure, used prior to the introduction of crankcase emission control systems. The tube, which was connected into the crankcase and suspended a few inches from the ground, depended on venturi action to create a partial vacuum as the vehicle moved. The method was ineffective below about 20 mph [32 km/h].

road load A constant vehicle speed on a level road.

rocker arm In an I-head engine, a device that rocks on a shaft (or pivots on a stud) as the cam moves the pushrod, causing a valve to open.

rod bearing In an engine, the bearing in the connecting rod in which a crankpin of the crankshaft rotates. Also called a *connecting-rod bearing*.

rod big end The end of the connecting rod that attaches around the crankpin.

rod bolts The bolts used to attach the cap to the connecting rod.

rod small end The end of the connecting rod through which a piston pin passes to connect the piston to the connecting rod.

roll bar A curved protective bar mounted over the driver's head in an off-road vehicle, high-performance car, or race car.

rolling radius The height of a tire, measured from the center of the rear axle to the ground.

RON Abbreviation for **research octane number.**

room temperature 68 to 72°F [20 to 22°C].

rotary Term describing the motion of a part that continually rotates or turns.

rotary-combustion engine See **Wankel engine.**

rotary-valve steering gear A type of power-steering gear.

rotor A revolving part of a machine, such as an alternator rotor, disk-brake rotor, distributor rotor, or Wankel-engine rotor.

rotor oil pump A type of oil pump in which a pair of rotors, one inside the other, produce the pressure required to circulate oil to engine parts.

rpm Abbreviation for *revolutions per minute,* a measure of rotational speed.

ruler A graduated straightedge used for measuring distances, usually up to 1 ft.

run-on See **dieseling.**

runout Wobble.

S

SA Designation for lubricating oil that is acceptable for use in engines operated under the mildest conditions.

SAE Abbreviation for *Society of Automotive Engineers.* Used to indicate a grade or weight of oil measured according to Society of Automotive Engineers standards.

safety Freedom from injury or danger.

safety rim A wheel rim with a hump on the inner edge of the ledge on which the tire bead rides. The hump helps hold the tire on the rim in case of a blowout.

sag A momentary decrease in acceleration rate; does not occur immediately after throttle application (as in a hesitation), but after the vehicle has acquired some speed.

sanitary Performance term used to describe a very clean or exceptionally well prepared racing or show car.

SB Designation for lubricating oil that is acceptable for minimum-duty engines operated under mild conditions.

SC Designation for lubricating oil that meets requirements for use in the gasoline engines in 1964 to 1967 passenger cars and trucks.

scavenging The displacement of exhaust gas from the combustion chamber by fresh air.

SCCA Abbreviation for *Sports Car Club of America.*

schematic A pictorial representation, most often in the form of a line drawing. A systematic positioning of components and their relationship to each other or to the overall function.

Schrader valve A spring-loaded valve through which a connection can be made to a refrigeration system; also used in tires.

scope Short for **oscilloscope.**

scored Scratched or grooved, as a cylinder wall may be scored by abrasive particles moved up and down by the piston rings.

scraper A device used in engine service to scrape carbon from the engine block, pistons, or other parts.

scraper ring On a piston, an oil-control ring designed to scrape excess oil back down the cylinder and into the crankcase.

screens Pieces of fine-mesh metal fabric; used to prevent solid particles from circulating through any liquid or vapor system and damaging vital moving parts. In an air conditioner, screens are located in the receiver-dehydrator, thermostatic expansion valve, and compressor.

screw A metal fastener with threads that can be turned into a threaded hole, usually with a screwdriver. There are many different types and sizes of screws.

screwdriver A hand tool used to loosen or tighten screws.

scuffing A type of wear in which there is a transfer of material between parts moving against each other; shows up as pits or grooves in the mating surfaces.

SD Designation for lubricating oil that meets requirements for use in the gasoline engines in 1968 to 1971 passenger cars and some trucks.

SDV Abbreviation for *spark delay valve;* a calibrated restrictor in the vacuum-advance hose which delays the vacuum spark advance.

SE Designation for lubricating oil that meets requirements for use in the gasoline engines in 1972 and later cars, and in certain 1971 passenger cars and trucks.

seal A material, shaped around a shaft, used to close off the operating compartment of the shaft, preventing oil leakage.

sealed-beam headlight A headlight that contains the filament, reflector, and lens in a single sealed unit.

sealer A thick, tacky compound, usually spread with a brush, which may be used as a gasket or sealant, to seal small openings or surface irregularities.

seat The surface upon which another part rests, as a valve seat. Also, to wear into a good fit; for example, new piston rings seat after a few miles of driving.

seat adjuster A device that permits forward and backward (and sometimes upward and downward) movement of the front seat of a vehicle.

secondary air Air that is pumped to thermal reactors, catalytic converters, exhaust manifolds, or the cylinder-head exhaust ports, to promote the chemical reactions that reduce exhaust-gas pollutants.

secondary available voltage Voltage that is available for firing the spark plug.

secondary circuit The high-voltage circuit of the ignition system; consists of the coil, rotor, distributor cap, spark-plug cables, and spark plugs.

section modulus A measure of the strength of the car-frame side rails; depends on the cross-sectional area and shape of the rails.

sector A gear that is not a complete circle. Specifically, the gear sector on the pitman shaft, in many steering systems.

segments The copper bars of a commutator.

self-adjuster A mechanism used on drum brakes; compensates for shoe wear by automatically keeping the shoe adjusted close to the drum.

self-discharge Chemical activity in the battery which causes the battery to discharge even though it is furnishing no current.

self-induction The inducing of a voltage in a current-carrying coil of wire because the current in that wire is changing.

self-locking screw A screw that locks itself in place, without the use of a separate nut or lockwasher.

self-tapping screw A screw that cuts its own threads as it is turned into an unthreaded hole.

semiconductor A material that acts as an insulator under some conditions and as a conductor under other conditions.

semiconductor ignition system See **electronic ignition system.**

semifloating rear axle An axle that supports the weight of the vehicle on the axle shaft in addition to transmitting driving forces to the rear wheels.

sensible heat Heat which can be felt or measured with a thermometer. Sensible heat changes the temperature of a substance, but not its state.

separator A thin sheet of wood, rubber, or glass mat that is placed between positive and negative plates in a battery cell to insulate them from each other.

series circuit An electric circuit in which the devices are connected end to end, positive terminal to negative terminal. The same current flows through all the devices in the circuit.

series-parallel system A starting system using two batteries, connected differently for different functions. For example, a system with a 24-volt starting motor, two 12-volt batteries, and a 12-volt alternator. For starting, the two batteries are connected in series to produce 24 volts; for charging, they are connected in parallel to produce 12 volts.

service manual A book published annually by each vehicle manufacturer, listing the specifications and service procedures for each make and model of vehicle. Also called a *shop manual*.

service rating A designation that indicates the type of service for which an engine lubricating oil is best suited. See **SA, SB, SC, SD,** and **SE.**

servo A device in a hydraulic system that converts hydraulic pressure to mechanical movement. Consists of a piston which moves in a cylinder as hydraulic pressure acts on it.

setscrew A type of metal fastener that holds a collar or gear on a shaft when its point is turned down into the shaft.

set up Performance term meaning to prepare a car for racing.

severe ring A piston ring which exerts relatively high pressure against the cylinder walls, sometimes by use of an expander spring located behind the ring; a ring that can be used in an engine with excessive cylinder wear.

shackle The swinging support by which one end of a leaf spring is attached to the car frame.

shaved Performance term for a vehicle from which the body chrome and hardware have been removed. Also used in referring to a cylinder head that has been resurfaced to increase the cylinder compression ratio.

shift lever The lever used to change gears in a transmission. Also, the lever on the starting motor which moves the drive pinion into or out of mesh with the flywheel teeth.

shift valve In an automatic transmission, a valve that moves to produce the shifts from one gear ratio to another.

shim A slotted strip of metal used as a spacer to adjust the front-end alignment on many cars; also used to make small corrections in the position of body sheet metal and other parts.

shimmy Rapid oscillation. In wheel shimmy, for example, the front wheel turns in and out alternately and rapidly; this causes the front end of the car to oscillate, or shimmy.

shim stock Sheets of metal of accurately known thicknesses which can be cut into strips and used to measure or correct clearances.

shock absorber A device placed at each vehicle wheel to regulate spring rebound and compression.

shoe In the brake system, a metal plate that supports the brake lining and absorbs and transmits braking forces.

shop layout The locations of aisles, work areas, machine tools, etc., in a shop.

short-arm, long-arm (SALA) suspension Name given to the conventional front-suspension system, which uses a short upper control arm and a longer lower control arm.

short circuit A defect in an electric circuit which permits current to take a short path, or circuit, instead of following the desired path.

shrink fit A tight fit of one part into another, achieved by heating or cooling one part and then assembling it to the other part. A heated part will shrink on cooling to provide the tight fit; a cooled part will expand on warming to provide the tight fit.

shroud A hood placed around an engine fan to improve fan action.

shunt A parallel connection or circuit.

side clearance The clearance between the sides of moving parts when the sides do not serve as load-carrying surfaces.

sight glass In a car air conditioner, a viewing glass or window set in the refrigerant line, usually in the top of the receiver-dehydrator; the sight glass allows a visual check of the refrigerant passing from the receiver to the evaporator.

single-chamber capacity In a Wankel engine, a measurement of the displacement or maximum volume of the rotor chamber.

single-overhead-camshaft (SOHC) engine An engine in which a single camshaft is mounted over each cylinder head, instead of inside the cylinder block.

skid control A device that operates to prevent wheel lockup during braking and, thus, to prevent skidding.

slick Performance term for a smooth, treadless racing tire.

slip joint In the power train, a variable-length connection that permits the drive shaft to change its effective length.

slip rings In an alternator, the rings that form a rotating connection between the armature windings and the brushes.

sludge An accumulation of water, dirt, and oil in the oil pan; sludge is very viscous and tends to reduce lubrication.

smog A term coined from the words *sm*oke and f*og*. First applied to the foglike layer that hangs in the air under certain atmospheric conditions; now generally used to describe any condition of dirty air and/or fumes or smoke. Smog is compounded from smoke, moisture, and numerous chemicals which are produced by combustion.

smoke Small gasborne or airborne particles, exclusive of water vapor, that result from combustion; such particles emitted by an engine into the atmosphere in sufficient quantity to be observable.

smoke in exhaust A visible blue or black substance often present in the automotive exhaust. A blue color indicates excessive oil in the combustion chamber; black indicates excessive fuel in the air-fuel mixture.

snap ring A metal fastener, available in two types: the *external* snap ring fits into a groove in a shaft; the *internal* snap ring fits into a groove in a housing. Snap rings must be installed and removed with special snap-ring pliers.

socket wrench A wrench that fits entirely over or around the head of a bolt.

sodium-cooled valve A hollow valve partially filled with metallic sodium. The sodium transfers heat (by convection) from the hot head end of the valve to the stem, thereby reducing the valve temperature.

SOHC See **single-overhead-camshaft engine.**

soldering Joining pieces of metal with solder, flux, and heat.

solenoid An electromechanical device which, when connected to an electrical source such as a battery, produces a mechanical movement. This movement can be used to control a valve or to produce other movements.

solenoid bypass valve In the air conditioner, a valve placed in a bypass line between the condenser and compressor, and operated by a solenoid. When the valve is open, refrigerant can bypass between the condenser and compressor, so no refrigeration takes place. When the valve is closed, the system cools normally.

solenoid relay A relay that connects a solenoid to a current source when its contacts close; specifically, the starting-motor solenoid relay.

solenoid switch A switch that is opened and closed electromagnetically, by the movement of a solenoid core. Usually, the core also causes a mechanical action, such as the movement of a drive pinion into mesh with flywheel teeth for cranking.

solid-state regulator An alternator regulator encapsulated in a plastic material and mounted in the alternator.

solvent A petroleum product of low volatility used in the cleaning of engine and vehicle parts.

solvent tank In the shop, a tank of cleaning fluid, in which most parts are brushed and washed clean.

south pole The pole at which magnetic lines of force enter a magnet.

spark advance See **advance.**

spark decel valve A vacuum-actuated valve, located between the carburetor and distributor, which advances the spark during deceleration to reduce emissions. (Should not be confused with the spark-delay valve.)

spark duration The length of time a spark is established across a spark gap, or the length of time current flows in a spark gap.

spark knock See **detonation.**

spark line Part of the oscilloscope pattern of the ignition secondary circuit; the spark line shows the voltage required to sustain the spark at the spark plug, and the number of distributor degrees through which the spark exists.

spark plug A device that screws into the cylinder head of an engine; provides a spark to ignite the compressed air-fuel mixture in the combustion chamber.

spark-plug heat range The distance heat must travel from the center electrode to reach the outer shell of the spark plug and enter the cylinder head.

spark test A quick check of the ignition system; made by holding the metal spark-plug end of a spark-plug cable about $3/16$ inch [4.76 mm] from the cylinder head, or block; cranking the engine; and checking for the existence and intensity of a spark.

specifications Information, provided by the manufacturer, that describes each automotive system and its components, operation, and clearances. Also, the service procedures that must be followed for a system to operate properly.

specific gravity The weight per unit volume of a substance as compared with the weight per unit volume of water.

specific heat The quantity of heat (in Btus) required to change the temperature of 1 lb of a substance by 1°F.

specs Short for **specifications.**

speed The rate of motion; for vehicles, measured in miles per hour or kilometers per hour.

speedometer An instrument that indicates vehicle speed; usually driven from the transmission.

speed shift Performance term for a shift of gears in a manual transmission without releasing the accelerator.

splash-feed oil system A type of engine lubricating system in which oil is splashed onto moving engine parts.

splines Slots or grooves cut in a shaft or bore; splines on a shaft are matched to splines in a bore, to ensure that two parts turn together.

spongy pedal Term applied to the feel of a brake pedal when air is trapped in the hydraulic system.

spool valve A rod with indented sections; used to control oil flow in automatic transmissions.

sprag In an automatic transmission, a one-way clutch; a clutch that can transmit power in one direction, but not in the other.

spring A device that changes shape under stress or pressure, but returns to its original shape when the stress or pressure is removed; the component of the automotive suspension system that absorbs road shocks by flexing and twisting.

spring rate The load required to move a spring or a suspended wheel a given distance; indicates the softness or firmness of a given spring or suspension.

spring retainer In the valve train, the piece of metal that holds the spring in place, and is itself locked in place by the valve-spring-retainer lock.

spring shackle See **shackle**.

sprung weight That part of the car which is supported on springs (includes the engine, frame, and body).

spur gear A gear in which the teeth are parallel to the center line of the gear.

square engine An engine in which the bore and stroke are equal.

squeak A high-pitched noise of short duration.

squeal A continuous, high-pitched, low-volume noise.

squish The action in some combustion chambers in which the last part of the compressed air-fuel mixture is pushed, or squirted, out of a decreasing space between the piston and cylinder head.

stabilizer bar An interconnecting shaft between the two lower suspension arms; reduces body roll on turns.

stacked pattern See **raster pattern**.

stacks Performance term for short, tubular carburetor intake pipes; also, for short, individual exhaust pipes.

stall test A starting-motor test in which the current draw is measured with the motor stalled.

standpipe assembly See **vapor-liquid separator**.

starter See **starting motor**.

starting motor The electric motor that cranks the engine, or turns the crankshaft, for starting.

starting-motor drive The drive mechanism and gear on the end of the starting-motor armature shaft; used to couple the starting motor to, and disengage it from, the flywheel ring-gear teeth.

static balance The balance of an object while it is not moving.

static friction The friction between two bodies at rest.

stator In the torque converter, a third member (in addition to the turbine and pump) which changes the direction of fluid flow under certain operating conditions (when the stator is stationary). In an alternator, the assembly that includes the stationary conductors.

steam cleaner A machine used for cleaning large parts with a spray of steam, often mixed with soap.

steam engine An external-combustion engine operated by steam generated in a boiler.

steering-and-ignition lock A device that locks the ignition switch in the OFF position and locks the steering wheel so it cannot be turned.

steering arm The arm, attached to the steering knuckle, that turns the knuckle and wheel for steering.

steering axis The center line of the ball joints in a front-suspension system.

steering-axis inclination The inward tilt of the steering axis from the vertical.

steering-column shift An arrangement in which the transmission shift lever is mounted on the steering column.

steering gear That part of the steering system that is located at the lower end of the steering shaft; carries the rotary motion of the steering wheel to the car wheels for steering.

steering geometry See **toe-out on turns.**

steering kickback Sharp and rapid movements of the steering wheel as the front wheels encounter obstructions in the road; the shocks of these encounters "kick back" to the steering wheel.

steering knuckle The front-wheel spindle which is supported by upper and lower ball joints and by the wheel; the part on which a front wheel is mounted, and which is turned for steering.

steering shaft The shaft extending from the steering gear to the steering wheel.

steering system The mechanism that enables the driver to turn the wheels for changing the direction of vehicle movement.

steering wheel The wheel, at the top of the steering shaft, which is used by the driver to guide, or steer, the car.

stepped feeler gauge A feeler gauge which has a thin tip, of a known dimension, and is thicker along the rest of the gauge; a "go no-go" feeler gauge.

Stirling engine A type of internal-combustion engine in which the piston is moved by changes in the pressure of a working gas that is alternately heated and cooled.

stoplights Lights, at the rear of a vehicle, which indicate that the brakes are being applied by the driver to slow or stop the vehicle.

stoplight switch The switch that turns the stoplights on and off as the brakes are applied and released.

storage battery A device that changes chemical energy into electrical energy; that part of the electric system which acts as a reservoir for electric energy, storing it in chemical form.

stratified charge In a gasoline-fueled spark-ignition engine, an air-fuel charge with a small layer of very rich air-fuel mixture; the rich layer is ignited first, after which ignition spreads to the leaner mixture filling the rest of the combustion chamber. The diesel engine is a stratified-charge engine.

streamlining The shaping of a car body or truck cab so that it minimizes air resistance and can be moved through the air with less energy.

stroke In an engine cylinder, the distance that the piston moves in traveling from BDC to TDC or from TDC to BDC.

stroker kit Performance term for a special crankshaft-and-connecting-rod kit designed to increase the displacement of an engine by lengthening the stroke of the piston.

strut A bar that connects the lower control arm to the car frame; used when the lower control arm is of the type that is attached to the frame at only one point. Also called a *brake reaction rod*.

stud A headless bolt that is threaded on both ends.

stud extractor A special tool used to remove a broken stud or bolt.

stumble A condition related to vehicle driveability; the tendency of an engine to falter, and then catch, resulting in a noticeable hesitation felt by the driver. A momentary abrupt deceleration during an acceleration.

substrate In a catalytic converter, the supporting structure to which the catalyst is applied; usually made of ceramic. Two types of substrate used in catalytic converters are the monolithic or one-piece substrate and the bead- or pellet-type substrate.

suction line In an air conditioner, the tube that connects the evaporator outlet and the compressor inlet. Low-pressure refrigerant vapor flows through this line.

suction pressure The pressure at the air-conditioner compressor inlet; the compressor intake pressure, as indicated by a gauge set.

suction side That portion of the refrigeration system under low pressure; extends from the thermostatic expansion valve to the compressor inlet.

suction throttling valve In an air conditioner, a valve located between the evaporator and the compressor; controls the temperature of the air flowing from the evaporator, to prevent freezing of moisture on the evaporator.

sulfation The lead sulfate that forms on battery plates as a result of the battery action that produces electric current.

sulfuric acid See **electrolyte.**

sulfur oxides (SO_x) Acids that can form in small amounts as the result of a reaction between hot exhaust gas and the catalyst in a catalytic converter.

sun gear In a planetary-gear system, the center gear that meshes with the planet pinions.

supercharger In the intake system of the engine, a device that pressurizes the ingoing air-fuel mixture. This increases the amount of mixture delivered to the cylinders and thus increases the engine output. If the supercharger is driven by the engine exhaust gas, it is called a **turbocharger.**

superheated vapor Refrigerant vapor at a temperature that is higher than its boiling point for a given pressure.

superimposed pattern On an oscilloscope, a pattern showing the ignition voltages one on top of the other, so that only a single trace, and variations from it, can be seen.

surface grinder A grinder used to resurface flat surfaces, such as cylinder heads.

surface ignition Ignition of the air-fuel mixture, in the combustion chamber, by hot metal surfaces or heated particles of carbon.

surge Condition in which the engine speed increases and decreases slightly but perceptibly, in spite of the fact that the driver has not changed the throttle position.

suspension The system of springs and other parts which supports the upper part of a vehicle on its axles and wheels.

suspension arm In the front suspension, one of the arms pivoted on the frame at one end, and on the steering-knuckle support at the other end.

S/V ratio The ratio of the surface area S of the combustion chamber to its volume V, with the piston at TDC. Often used as a comparative indicator of hydrocarbon emission levels from an engine.

sway bar See **stabilizer bar.**

switch A device that opens and closes an electric circuit.

synchro Performance term for synchromesh transmission.

synchromesh transmission A transmission with a built-in device that automatically matches the rotating speeds of the transmission gears, thereby eliminating the need for "double-clutching."

synchronize To make two or more events or operations occur at the same time or at the same speed.

synchronizer A device in the transmission that synchronizes gears about to be meshed, so that there will not be any gear clash.

T

tachometer A device for measuring engine speed, or revolutions per minute.

taillights Steady-burning low-intensity lights used on the rear of a vehicle.

tank unit The part of the fuel-indicating system that is mounted in the fuel tank.

tap A tool used for cutting threads in a hole.

taper A gradual reduction in the width of a shaft or hole; in an engine cylinder, uneven wear, more pronounced at the top than at the bottom.

tappet See **valve lifter**.

taxable horsepower The power of an engine, as calculated by a formula that provides a comparison of engines on a uniform basis. The formula is used in some localities for licensing vehicles.

TCS See **transmission-controlled spark system**.

TDC Abbreviation for **top dead center**.

TEL Abbreviation for **tetraethyl lead**.

temperature The measure of heat intensity or concentration, in degrees. Temperature is not a measure of heat quantity.

temperature gauge A gauge that indicates, to the driver, the temperature of the coolant in the engine cooling system.

temperature indicator See **temperature gauge**.

temperature-sending unit A device, in contact with the engine coolant, whose electrical resistance changes as the coolant temperature increases or decreases; these changes control the movement of the indicator needle of the temperature gauge.

tetraethyl lead A chemical which, when added to engine fuel, increases its octane rating, or reduces its knocking tendency. Also called *ethyl*.

thermactor See **air-injection system**.

thermal Of or pertaining to heat.

thermal-conductivity gas analyzer The conventional exhaust-gas analyzer, used in service shops for many years to check and adjust the carburetor air-fuel mixtures.

thermal efficiency Ratio of the energy output of an engine to the energy in the fuel required to produce that output.

thermal reactor A large exhaust manifold in which unburned exhaust-gas hydrocarbons and carbon monoxide react with oxygen so that the pollutants burn up almost completely. It is simple and durable, but must operate at very high temperatures. Used on the Mazda car with the Wankel engine.

thermistor A heat-sensing device with a negative temperature coefficient of resistance; that is, as its temperature increases, its electrical resistance decreases. Used as the sensing device for engine-temperature indicating instruments.

thermometer An instrument which measures heat intensity (temperature) via the thermal expansion of a liquid.

thermostat A device for the automatic regulation of temperature; usually contains a temperature-sensitive element that expands or contracts to open or close off the flow of air, a gas, or a liquid.

thermostatically controlled air cleaner An air cleaner in which a thermostat controls the preheating of intake air.

thermostatic expansion valve Component of a refrigeration system that controls the rate of refrigerant flow to the evaporator. Commonly called the *TE valve, TXV,* or simply the *expansion valve.*

thermostatic gauge An indicating device (for fuel quantity, oil pressure, engine temperature, etc.) that contains a thermostatic blade or blades.

thermostatic switch An adjustable component used in a cycling-clutch air conditioner. Engages and disengages the compressor to prevent water from freezing on the evaporator core, and control the temperature of air flowing from the evaporator.

thermostatic vacuum switch A temperature-sensing device screwed into the coolant system; connects full manifold vacuum to the distributor when its critical temperature is reached. The resultant spark advance causes an increase in engine rpm, which cools the engine.

thread chaser A device, similar to a die, that is used to clean threads.

thread class A designation indicating the closeness of fit between a pair of threaded parts, such as a nut and bolt.

threaded insert A threaded coil that is used to restore the original thread size to a hole with damaged threads; the hole is drilled oversize and tapped, and the insert is threaded into the tapped hole.

thread series A designation indicating the pitch, or number of threads per inch, on a threaded part.

three-mode cycle A quick test procedure used over the past several years to study the causes of high emissions and to compare different types of testers; consists of taking readings at idle speed and at 2,000 rpm, and maximum readings on deceleration. The test can be performed on a dynamometer under load, or in a service area without load.

three-phase Designating a combination of three interconnected ac circuits in which the alternations are one-third of a cycle apart.

throat Performance term for a carburetor venturi.

throttle A disk valve in the carburetor base that pivots in response to accelerator-pedal position; allows the driver to regulate the volume of air-fuel mixture entering the intake manifold, thereby controlling the engine speed. Also called the *throttle plate* or **throttle valve.**

throttle-return check Same as **dashpot.**

throttle solenoid positioner An electric solenoid which holds the throttle plate open (hot-idle position), but also permits the throttle plate to close completely when the ignition is turned off, to prevent "dieseling."

throttle valve A round disk valve in the throttle body of the carburetor; can be turned to admit more or less air, thereby controlling engine speed.

throw a rod In an engine, to have a loose, knocking connecting-rod bearing, or a broken connecting rod that has been forced through the cylinder block or oil pan.

throwout bearing In the clutch, the bearing that can be moved in to the release levers by clutch-pedal action, to cause declutching—disengaging the engine crankshaft from the transmission.

thrust bearing In the engine, the main bearing that has thrust faces which prevent excessive end play, or forward and backward movement of the crankshaft.

tie-rod end A socket and a ball stud in a housing. They rotate and tilt to transmit steering action under all conditions.

tie rods In the steering system, the rods that link the pitman arm to the steering-knuckle arms; small steel components that connect the front wheels to the steering mechanism.

tilt steering wheel A type of steering wheel that can be tilted at various angles, through a flex joint in the steering shaft.

timing In an engine, delivery of the ignition spark or operation of the valves (in relation to the piston position) for the power stroke. See **ignition timing** and **valve timing.**

timing chain A chain that is driven by a sprocket on the crankshaft and that drives the sprocket on the camshaft.

timing gear A gear on the crankshaft; drives the camshaft by meshing with a gear on its end.

timing light A light that can be connected to the ignition system to flash each time the No. 1 spark plug fires; used for adjusting the timing of the ignition spark.

tire The casing-and-tread assembly (with or without a tube) that is mounted on a car wheel to provide pneumatically cushioned contact and traction with the road.

tire carcass The plies that constitute the underbody of the tire; the "skeleton" over which the rubber of the sidewalls and the thicker tread area are molded.

tire casing Layers of cord, called *plies,* shaped in a tire form and impregnated with rubber, to which the tread is applied.

tire rotation The interchanging of the running locations of the tires on a car, to minimize noise and to equalize tire wear.

tire tread See **tread.**

tire tube An inflatable rubber device mounted inside some tires to contain air at sufficient pressure to inflate the casing and support the vehicle weight.

tire-wear indicator Small strips of rubber molded into the bottom of the tire tread grooves; they appear as narrow strips of smooth rubber across the tire when the tread depth decreases to $\frac{1}{16}$ in [1.59 mm].

toe-in The amount, in inches or millimeters, by which the front of a front wheel points inward.

toe-out on turns The difference between the angles each of the front wheels makes with the car frame, during turns. On a turn, the inner wheel turns, or toes out, more. Also called the **steering geometry.**

top dead center The piston position when the piston has reached the upper limit of its travel in the cylinder, and the center line of the connecting rod is parallel to the cylinder walls.

torque Turning or twisting effort; usually measured in pound-feet or kilogram-meters. Also, a turning force such as that required to tighten a connection.

torque converter In an automatic transmission, a fluid coupling which incorporates a stator to permit a torque increase.

Torqueflite An automatic transmission used on Chrysler-manufactured cars; has three forward speeds and reverse.

torque test A starting-motor test in which both the torque developed and the current drawn are measured while the specified voltage is applied.

torque wrench A wrench that indicates the amount of torque being applied with the wrench.

torsional balancer See **vibration damper.**

torsional vibration Rotary vibration that causes a twist-untwist action on a rotating shaft, so that a part of the shaft repeatedly moves ahead of, or lags behind, the remainder of the shaft; for example, the action of a crankshaft responding to the cylinder firing impulses.

torsion-bar spring A long, straight bar that is fastened to the vehicle frame at one end and to a suspension part at the other. Spring action is produced by a twisting of the bar.

torsion-bar steering gear A rotary-valve power-steering gear.

tracking Rear wheels following directly behind (in the tracks of) the front wheels.

tractive effort The force available at the road surface in contact with the driving wheels of a truck. It is determined by engine torque, transmission ratio, axle ratio, tire size, and frictional losses in the driveline. *Rim pull* is also known as tractive effort.

tractor A truck or comparatively short, wheelbase vehicle used for pulling a semitrailer or trailer; a self-propelled vehicle having tracks or wheels, used for pulling agricultural implements or mounting construction implements.

tractor breakaway valve A valve that couples the tractor and trailer emergency-brake systems. Provides air to the trailer emergency-brake system for normal operating conditions. If the trailer brake system fails, the breakaway valve automatically seals off the tractor braking system and activates the trailer emergency brake.

tramp Up-and-down motion (hopping) of the front wheels at higher speeds, due to unbalanced wheels or excessive wheel runout. Also called *high-speed shimmy*.

transducer Any device which converts an input signal of one form into an output signal of a different form. For example, the automobile horn converts an electric signal to sound.

transistor An electronic device that can be used as an electric switch; used to replace the contact points in electronic ignition systems.

transmission An assembly of gears that provides the different gear ratios, as well as neutral and reverse, through which engine power is transmitted to the differential to rotate the drive wheels.

transmission-controlled spark (TCS) system A General Motors NO_x exhaust-emission control system; makes use of the transmission-gear position to allow distributor vacuum advance in high gear only.

transmission-oil cooler A small radiator, either mounted separately or as part of the engine radiator, which cools the transmission fluid.

transmission-regulated spark (TRS) system A Ford exhaust-emission control system, similar to the General Motors transmission-controlled spark system; allows distributor vacuum advance in high gear only.

tread The part of the tire that contacts the road. It is the thickest part of the tire, and is cut with grooves to provide traction for driving and stopping.

trouble diagnosis The detective work necessary to find the cause of a trouble.

TRS See **transmission-regulated spark system.**

truck Any motor vehicle, primarily designed for the transportation of property, which carries the load on its own wheels.

truck tractor Any motor vehicle designed primarily for pulling truck trailers and constructed so as to carry part of the weight and load of a semitrailer.

tube-and-fin radiator core A type of radiator core consisting of tubes to which cooling fins are attached; coolant flows through the tubes, between the upper and lower radiator tanks.

tubeless tire A tire that holds air without the use of a tube.

tuned intake system An engine air intake system in which the manifold has the proper length and volume to introduce a ramjet or supercharging effect.

tune-up A procedure for inspecting, testing, and adjusting an engine, and replacing any worn parts, to restore the engine to its best performance.

turbine A device that produces rotary motion as a result of fluid pressure. Also, the driven member in a torque converter.

turbine engine An engine in which gas pressure created by combustion is used to spin a turbine and, through gears, move a vehicle.

turbocharger A supercharger driven by the engine exhaust gas.

Turbo Hydra-Matic An automatic transmission built by General Motors and used on many models of cars and light trucks; has three forward speeds and reverse.

turbulence The state of being violently disturbed. In the engine, the rapid swirling motion imparted to the air-fuel mixture entering a cylinder.

turn signal See **directional signal.**

TVS Abbreviation for **thermostatic vacuum switch.**

twenty-five-ampere rate A battery rating; the length of time a battery can deliver 25 A before the cell voltage drops to 1.75 volts, starting with the electrolyte at 80°F [26.7°C].

twenty-hour rate A battery rating; the amount of current a battery can deliver for 20 hours before the cell voltage drops below 1.75 volts, starting with an electrolyte temperature of 80°F [26.7°C].

twin I-beam A type of front-suspension system used on some trucks.

twist drill A conventional drill bit.

two-barrel carburetor A carburetor with two throttle valves.

two cycle Short for **two-stroke cycle.**

two-disk clutch A clutch with two friction disks for additional holding power; used in heavy-duty equipment.

two-stroke cycle The two piston strokes during which fuel intake, compression, combustion, and exhaust take place in a two-stroke-cycle engine.

U

U bolt An iron rod with threads on both ends, bent into the shape of a U and fitted with a nut at each end.

under-dash unit The "hang-on" type air-conditioning system installed under the dash after a vehicle leaves the factory. All air outlets are in the evaporator case, and the system normally uses only recirculated air. The discharge air temperature is controlled by a cycling thermostatic expansion switch or a suction throttling valve.

unit An assembly or device that can perform its function only if it is not further divided into its components.

unit distributor A General Motors ignition distributor that uses a magnetic pickup coil and timer core instead of points and a condenser. The ignition coil is assembled into the distributor as a unit.

unitized construction A type of automotive construction in which the frame and body parts are welded together to form a single unit.

universal joint In the power train, a jointed connection in the drive shaft that permits the driving angle to change.

unleaded gasoline Gasoline to which no lead compounds have been intentionally added. Gasoline that contains 0.05 g or less of lead per gallon; required by law to be used in 1975 and later vehicles equipped with catalytic converters.

unloader A device linked to the throttle valve; opens the choke valve when the throttle is moved to the wide-open position.

upper beam A headlight beam intended primarily for distant illumination, not for use when meeting or following other vehicles.

upshift To shift a transmission into a higher gear.

unsprung weight The weight of that part of the car which is not supported on springs; for example, the wheels and tires.

USAC Abbreviation for *United States Auto Club*.

V

vacuum Negative gauge pressure, or a pressure less than atmospheric pressure. Vacuum can be measured in psi, but is usually measured in inches or millimeters of mercury (Hg); a reading of 30 inches [762 mm] Hg would indicate a perfect vacuum.

vacuum advance The advancing (or retarding) of ignition timing by changes in intake-manifold vacuum, which reflect throttle opening and engine load. Also, a mechanism on the ignition distributor that uses intake-manifold vacuum to advance the timing of the spark to the spark plugs.

vacuum-advance control Any type of NO_x emission control system designed to allow vacuum advance only during certain modes of engine and vehicle operation.

vacuum-advance solenoid An electrically operated two-position valve which allows or denies intake-manifold vacuum to the distributor vacuum-advance unit.

vacuum-control temperature-sensing valve A valve that connects manifold vacuum to the distributor advance mechanism under hot-idle conditions.

vacuum gauge In automotive-engine service, a device that measures intake-manifold vacuum and thereby indicates actions of engine components.

vacuum modulator In automatic transmissions, a device that modulates, or changes, the main-line hydraulic pressure to meet changing engine loads.

vacuum motor A small motor, powered by intake-manifold

vacuum; used for jobs such as raising and lowering headlight doors.

vacuum power unit A device for operating accessory doors and valves, using vacuum as a source of power.

vacuum pump A mechanical device used to evacuate a system.

vacuum-suspended power brake A type of power brake in which both sides of the piston are subjected to vacuum; the piston is thus "suspended" in vacuum.

vacuum switch A switch that closes or opens its contacts in response to changing vacuum conditions.

valve A device that can be opened or closed to allow or stop the flow of a liquid or gas.

valve clearance The clearance between the rocker arm and the valve-stem tip in an overhead-valve engine; the clearance in the valve train when the valve is closed.

valve float A condition in which the engine valves do not close completely, or fail to close at the proper time.

valve grinding Refacing a valve in a valve-refacing machine.

valve guide A cylindrical part in the cylinder block or head in which a valve is assembled and in which it moves up and down.

valve-in-head engine See **I-head engine.**

valve lash Same as **valve clearance.**

valve lifter A cylindrical part of the engine which rests on a cam of the camshaft and is lifted, by cam action, so that the valve is opened. Also called a *lifter, tappet, valve tappet,* or *cam follower.*

valve-lifter foot The bottom end of the valve lifter; the part that rides on the cam lobe.

valve overlap The number of degrees of crankshaft rotation during which the intake and exhaust valves are open together.

valve rack Any storage container or holder which identifies the valves and keeps them in order after they are removed from the engine.

valve-refacing machine A machine for removing material from the seating face of a valve to true up the face.

valve rotator A device which rotates the valve slightly each time it opens; this causes deposits to be wiped off the valve face and stem, ensuring good heat transfer from the face to the valve seat.

valve seat The surface against which a valve comes to rest to provide a seal against leakage.

valve-seat inserts Metal rings inserted in valve seats (usually for exhaust valves); they are made of special metals able to withstand very high temperatures.

valve-seat recession The tendency for valves, in some engines run on unleaded gasoline, to contact the seat in such a way that the seat wears away, or recesses, into the cylinder head. Also known as *lash loss*.

valve spool A spool-shaped valve, such as in the power-steering unit.

valve-spring retainer The device on the valve stem that holds the valve spring in place.

valve-spring-retainer lock The device on the valve stem that locks the valve-spring retainer in place.

valve stem The long, thin section of the valve that fits in the valve guide.

valve-stem seal A device placed on or around the valve stem to reduce the amount of oil that can get on the stem, and then work its way down into the combustion chamber. Also called a *valve-stem shield*.

valve tappet See **valve lifter**.

valve timing The timing of the opening and closing of the valves in relation to the piston position.

valve train The valve-operating mechanism of an engine; includes all components from the camshaft to the valve.

vane A flat, extended surface that is moved around an axis by or in a fluid. Part of the internal revolving portion of an air-supply pump.

vapor A gas; any substance in the gaseous state, as distinguished from the liquid or solid state.

vapor-fuel separator Same as **vapor-liquid separator**.

vaporization A change of state from liquid to vapor or gas, by evaporation or boiling; a general term including both evaporation and boiling.

vapor lines Lines that carry refrigerant vapor. See **suction line, discharge line,** and **equalizer line.**

vapor-liquid separator A device in the evaporative emission control system; prevents liquid gasoline from traveling to the engine through the charcoal-canister vapor line.

vapor lock A condition in the fuel system in which gasoline vaporizes in the fuel line or fuel pump; bubbles of gasoline vapor restrict or prevent fuel delivery to the carburetor.

vapor-recovery system An evaporative emission control system that recovers gasoline vapor escaping from the fuel tank and carburetor float bowl. See **evaporation control system.**

vapor-return line A line from the fuel pump to the fuel tank; allows vapor that has formed in the fuel pump to return to the fuel tank.

vapor-saver system Same as **vapor-recovery system.**

variable-ratio power steering Power-steering system in which the response of the car wheels varies according to how much the steering wheel is turned.

V-block A metal block with an accurately machined V-shaped groove; used to support an armature or shaft while it is checked for roundness.

VDV Abbreviation for *vacuum-delay valve.*

vehicle identification number (VIN) The number assigned to each vehicle by its manufacturer, primarily for registration and identification purposes.

vehicle vapor recovery See **vapor-recovery system.**

V-8 engine An engine with two banks of four cylinders each, set at an angle to form a V.

V engine See **V-type engine.**

vent An opening through which air can leave an enclosed chamber.

ventilation The circulating of fresh air through any space, to replace impure air. The basis of crankcase ventilation systems.

venturi In the carburetor, a narrowed passageway or restriction which increases the velocity of air moving through it; produces the vacuum responsible for the discharge of gasoline from the fuel nozzle.

VI Abbreviation for **viscosity index.**

vibration A rapid back-and-forth motion; an oscillation.

vibration damper A device attached to the crankshaft of an engine to oppose crankshaft torsional vibration (that is, the twist-untwist actions of the crankshaft caused by the cylinder firing impulses). Also called a *harmonic balancer*.

VIN Abbreviation for **vehicle identification number.**

viscosity The resistance to flow exhibited by a liquid. A thick oil has greater viscosity than a thin oil.

viscosity index A number indicating how much the viscosity of an oil changes with heat.

viscosity rating An indicator of the viscosity of engine oil. There are separate ratings for winter driving and for summer driving. The winter grades are SAE5W, SAE10W, and SAE20W. The summer grades are SAE20, SAE30, SAE40, and SAE50. Many oils have multiple-viscosity ratings, as, for example, SAE10W-30.

viscous Thick; tending to resist flowing.

viscous friction The friction between layers of a liquid.

vise A gripping device; used to hold a part steady while it is being worked on.

volatile Evaporating readily. For example, Refrigerant-12 is volatile (evaporates quickly) at room temperature.

volatility A measure of the ease with which a liquid vaporizes; has a direct relationship to the flammability of a fuel.

voltage The force which causes electrons to flow in a conductor. The difference in electrical pressure (or potential) between two points in a circuit.

voltage drop The reduction (drop) in voltage across an electrical device or a circuit; due to the resistance of the device or circuit.

voltage potential The electrical pressure at a particular point.

voltage regulator A device that prevents excessive alternator or generator voltage by alternately inserting and removing a resistance in the field circuit.

voltmeter A device for measuring the potential difference (voltage) between two points, such as the terminals of a battery or alternator, or two points in an electric circuit.

volumetric efficiency The ratio of the amount of air-fuel mixture that actually enters an engine cylinder to the theoretical amount that could enter under ideal conditions.

V-type engine An engine with two banks or rows of cylinders, set at an angle to form a V.

vulcanizing A process of treating raw rubber with heat and pressure; the treatment forms the rubber and gives it toughness and flexibility.

VVR Abbreviation for *vehicle-vapor recovery*. See **vapor-recovery system.**

W

wail Performance term used to describe a vehicle that is accelerating at a high rate of speed.

Wankel engine A rotary engine in which a three-lobe rotor turns eccentrically in an oval chamber to produce power.

warning blinker See **hazard system.**

water jackets The spaces between the inner and outer shells of the cylinder block or head, through which coolant circulates.

water pump In the cooling system, the device that circulates coolant between the engine water jackets and the radiator.

wedge combustion chamber A combustion chamber resembling a wedge in shape.

weight distribution The percentage of a vehicle's total weight that rests on each axle.

weight, sprung See **sprung weight.**

weight, unsprung See **unsprung weight.**

welding The process of joining pieces of metal by fusing them together with heat.

wet-disk clutch A clutch in which the friction disk (or disks) is operated in a bath of oil.

wheel alignment A series of tests and adjustments to ensure that wheels and tires are properly positioned on the vehicle.

wheel balancer A device that checks a wheel-and-tire assembly either statically or dynamically, for balance.

wheelbase The distance between the center lines of the

front and rear axles. For trucks with tandem rear axles, the rear center line is considered to be midway between the two rear axles.

wheel cylinders In a hydraulic braking system, hydraulic cylinders located in the brake mechanisms at the wheels. Hydraulic pressure from the master cylinder causes the wheel cylinders to move the brake shoes into contact with the brake drums for braking.

wheelie Performance term for lifting the front wheel of a motorcycle (or the front wheels of a car) off the ground during acceleration.

wheel tramp Tendency for a wheel to move up and down so it repeatedly bears down hard, or "tramps," on the road. Sometimes called *high-speed shimmy.*

window regulator A device for opening and closing a window; usually operated by a crank.

windshield wiper A mechanism which moves a rubber blade back and forth to wipe the windshield; operated either by vacuum or electrically.

wire feeler gauge A set of round wires of known diameters; used to check clearances between electrical contacts, such as distributor points and spark-plug electrodes.

wiring harness A group of individually insulated wires, wrapped together to form a neat, easily installed bundle.

work The changing of the position of an object against an opposing force; measured in foot-pounds or meter-kilograms. The product of a force and the distance through which it acts.

worm Type of gear in which the teeth resemble threads; used on the lower end of the steering shaft.

WOT Abbreviation for *wide-open throttle.*

wrench A tool designed for tightening and loosening nuts and bolts.

wrist pin See **piston pin.**

Z

zip gun An air-powered cutting tool often used for work on vehicle exhaust systems.

Abbreviations used in the U.S. Customary System

British thermal unit	Btu
degree(s) Fahrenheit	°F
fluid ounce (capacity)	fl oz
foot	ft
square foot	sq ft *or* ft^2
cubic foot	cu ft *or* ft^3
foot-pound (foot-pound-force)	ft-lb (ft-lbf)
gallon	gal
horsepower	HP *or* hp
inch	in. *or* in
square inch	sq in *or* in^2
cubic inch	cu in *or* in^3
mile	mi
miles per gallon	mpg
miles per hour	mph
ounce (weight)	oz
pint	pt
pound (pound-force)	lb (lbf)
*pounds per square inch	psi
quart	qt
ton	T
yard	yd

*U.S. trade usage

Abbreviations used in the SI System

centi− (1/100)	c−
centimeter	cm
cubic centimeter	cm³ *or* cc *or* cu cm
degrees(s) Celsius	°C
gram	g
joule	J
kilo− (× 1000)	k−
kilogram	kg
kilometer	km
kilopascal	kPa
kilowatt	kW
*kilogram-meters	kg m
*kilogram-centimeters	kg cm
*kilograms per square centimeter	kg/cm²
liter	*written* "l" *or spelled*
meter	m
square meter	m²
cubic meter	m³
milli− (1/1000)	m−
millimeter	mm
newton	N
newton-meter	N m
pascal	Pa
(metric) ton	t *or* Mt
watt	W

*U.S. trade usage

Approximate U.S. Customary System and SI Conversions

LENGTH

Inches	×	25.4	= millimeters
Inches	×	2.54	= centimeters
Feet	×	30	= centimeters
Feet	×	0.3	= meters
Miles	×	1.6	= kilometers
Millimeters	×	0.040	= inches
Centimeters	×	0.400	= inches
Centimeters	×	0.033	= feet
Meters	×	40	= inches
Meters	×	3.3	= feet
Kilometers	×	0.62	= miles

AREA

Square inches	×	6.5	= square centimeters
Square feet	×	0.09	= square meters
Square centimeters	×	0.16	= square inches
Square meters	×	11	= square feet

VOLUME

Cubic inches	×	16.4	= cubic centimeters
Cubic feet	×	0.03	= cubic meters
Cubic centimeters	×	0.06	= cubic inches
Cubic meters	×	35	= cubic feet

CAPACITY (Liquid Measure)

Fluid ounces	×	29.57	= cubic centimeters
Pints	×	0.47	= liters
Quarts	×	0.95	= liters
Gallons	×	3.8	= liters
Cubic centimeters	×	0.0338	= fluid ounces
Liters	×	2.1	= pints
Liters	×	1.06	= quarts
Liters	×	0.26	= gallons

FUEL PERFORMANCE

Miles per gallon	×	0.425	= kilometers per liter
Kilometers per liter	×	2.353	= miles per gallon

MASS AND WEIGHT

Ounces	×	28	= grams
Ounces	×	0.028	= kilograms
Pounds	×	0.45	= kilograms
Tons	×	0.9	= metric tons
Grams	×	0.035	= ounces
Kilograms	×	35	= ounces
Kilograms	×	2.2	= pounds
Metric tons	×	1.1	= tons

FORCE

Pounds (pounds-force)	×	4.448	= newtons
Newtons	×	0.2248	= pounds (pounds-force)

ENERGY

Foot-pounds (-force)	×	1.356	= joules
Foot-pounds (-force)	×	0.138	= kilogram-meters
British thermal units	×	1055	= joules
Joules	×	0.737	= foot-pounds (-force)
Kilogram-meters	×	7.233	= foot-pounds (-force)

Joules	× 0.00095	= British thermal units

POWER

Horsepower	× 0.746	= kilowatts
Kilowatts	× 1.340	= horsepower

VELOCITY

Miles per hour	× 1.6	= kilometers per hour
Kilometers per hour	× 0.625	= miles per hour

PRESSURE

Pounds per square inch	× 6.895	= kilopascals
Pounds per square inch	× 0.07	= kilograms per square centimeter
Kilopascals	× 0.145	= pounds per square inch
Kilograms per square centimeter	× 14.22	= pounds per square inch

TORQUE

Pound-inches	× 1.15	= kilogram-centimeters

Pound-feet	× 13.8	= kilogram-centimeters
Pound-feet	× 0.138	= kilogram-meters
Pound-inches	× 0.113	= newton-meters
Pound-feet	× 1.356	= newton-meters
Kilogram-centimeters	× 0.87	= pound-inches
Kilogram-centimeters	× 0.0723	= pound-feet
Kilogram-meters	× 7.23	= pound-feet
Newton-meters	× 8.85	= pound-inches
Newton-meters	× 0.737	= pound-feet

TEMPERATURE

$5/9 \times$ (degrees Fahrenheit $-$ 32) = degrees Celsius

$(9/5 \times$ degrees Celsius$) + 32$ = degrees Fahrenheit